PROTOPLASMATOLOGIA

HANDBUCH DER PROTOPLASMAFORSCHUNG

BEGRÜNDET VON

L. V. HEILBRUNN · F. WEBER
PHILADELPHIA   GRAZ

HERAUSGEGEBEN VON

M. ALFERT · H. BAUER · C. V. HARDING · W. SANDRITTER · P. SITTE
BERKELEY   TÜBINGEN   ROCHESTER   FREIBURG I. BR.   FREIBURG I. BR.

MITHERAUSGEBER

J. BRACHET-BRUXELLES · H. G. CALLAN-ST. ANDREWS · R. COLLANDER-HELSINKI
K. DAN-TOKYO · E. FAURÉ-FREMIET-PARIS · A. FREY-WYSSLING-ZÜRICH
L. GEITLER-WIEN · K. HÖFLER-WIEN · M. H. JACOBS-PHILADELPHIA
N. KAMIYA-OSAKA · W. MENKE-KÖLN · A. MONROY-PALERMO
A. PISCHINGER-WIEN · J. RUNNSTRÖM-STOCKHOLM

BAND VI

KERN- UND ZELLTEILUNG

B

THE CHROMOSOME CYCLE

SPRINGER-VERLAG WIEN GMBH

THE CHROMOSOME CYCLE

BY

BERNARD JOHN and KENNETH R. LEWIS
BIRMINGHAM   OXFORD

WITH 45 FIGURES

SPRINGER-VERLAG WIEN GMBH

TITEL-NR. 8740

Protoplasmatologia
VI. Kern- und Zellteilung
B. The Chromosome Cycle

The Chromosome Cycle

By

Dr. BERNARD JOHN

Department of Genetics, The University, Birmingham, England

and

Dr. KENNETH R. LEWIS

Botany School, The University, Oxford, England

With 45 Figures

Contents

Introduction

The chromosome theory of mendelian heredity had a long and often un-easy development. The heated disputes which were common during the first part of this century arose partly from a lack of understanding between investigators who, though they studied the same phenomena, used different techniques. Lack of communication also contributed to the confusion. Thus, the important cytological contributions of the nineteenth century were performed and presented in ignorance of the crucial experiments of MENDEL. Consequently, when McCLUNG and others attempted a consistent synthesis opposition was experienced from both sides.

In retrospect we can see that the essential elements of the chromosome theory were already evident prior to the rediscovery of Mendelism. This theory has two principal features, one relating to genetic properties, the other to epigenetic effects. In regard to the former, the studies of STRAS-BURGER, VAN BENEDEN and others had shown that nuclei do not arise anew, that they produce daughter nuclei and do so by an accurate process of mitosis. It thus appeared that the chromosome complement was perpetuated unchanged during development. Further, the studies of HERTWIG on fertili-zation and of VAN BENEDEN and WEISMANN on meiosis showed that the nucleus and its chromosomes were implicated also in sexual heredity but that the rules governing their transmission in this connection were somewhat differ-ent. In fact, on the basis of this kind of evidence alone, WEISMANN com-pletely rejected all notions of Lamarckian inheritance and proposed that heredity depended on a substance with a definite chemical and molecular composition.

Actually, a demonstration of the epigenetic role of the nucleus, the second feature of the chromosome theory, was provided soon afterwards by BOVERI. Thus, even before the end of the nineteenth century there was evidence that the chromosomes controlled heredity on the one hand, and determined development on the other.

Now, to those who approached heredity from its material basis, it must have been obvious that the determinant was one thing and the determined was quite another, for how could the chromosome and the character be one and the same? However, this fundamental principle was far from obvious to the immaterialists who studied heredity as they saw it manifested in the phenotypic relations between parents and their offspring in sexual repro-duction. Indeed, even after the rediscovery of MENDEL's paper, there were still those who not only failed to make the distinction but failed to see the necessity for making it. Yet, clearly, the problem of development is largely one of filling "the vacuum between determinant and character" (DARLINGTON 1951).

Nowadays the chromosome theory can be presented in much greater detail and with utter confidence, but its two main features remain the same. However, while the role of the chromosomes in heredity and development has been appreciated for a long time, the manner in which they perform their genetic and epigenetic functions has become amenable to critical investigation only in recent years. There is, therefore, still an unmistakable tendency to think of chromosomes in terms of the discrete threads of cell division and, in keeping with this conception, the chromosome cycle is generally considered in relation to the microscopically visible changes in morphology which occur during the mechanically active phases of mitosis and meiosis.

Chromosome phenotype, however, changes not only during division but throughout the cell cycle. The changes which occur during interphase are, of course, scarcely revealed in morphological modifications of the restless "resting" nucleus. Consequently they are less obvious and correspondingly less amenable to investigation. This accounts for the concentration on the countable karyotype, with its visible properties of pairing and pycnosity, and the measurable movements of separation and segregation.

This approach towards the karyotype is as incomplete, and correspondingly misleading, as an attitude towards the virus based entirely on the structure of its infective phase for this, like the coiled chromosome, is but an abstraction from a continuous cycle of genesis and epigenesis. In fact, discrete, crystallizable viruses and compact, countable chromosomes both represent quiescent states associated with transmission.

We have explored the numerical and structural diversity of the karyotype within and between individuals and the mechanically significant variations of meiotic chromosome behaviour in two previously published monographs to this handbook, namely "The Chromosome Complement" (1968) and "The Meiotic System" (1965). In the present precis we are concerned principally with the auto- and hetero-synthetic activities of the chromosomes at different phases of the cell cycle. Some attention has, however, been paid to the chromosome as a visible organelle and to the dividing nucleus because the switch from metabolic to mechanical activity and back again is not made abruptly. Movement and synthesis overlap. What is more, certain metabolic events of interphase are meaningful only in relation to the subsequent division of the cell. Other contributors to volume VI (Kern- und Zellteilung) of this handbook deal also with particular aspects of chromosome behaviour which complement, supplement or extend the information presented here. Particularly pertinent are:

C = Endomitosis (1953), D = Riesenchromosomen (1962), G 1 = The Behavior of Centrioles and the Structure and Formation of the Achromatic Figure (1966), and G 2 = Chromosomenbewegung (in preparation).

The chromosome cycle has long been a subject of extensive research but the increased accuracy and specificity of modern methods have put recent investigations on a different plane from those of the past. The application of new and not wholly familiar techniques to a variety of materials inevitably means uncertainty, dispute and revision. In our view, therefore, it is premature to attempt an authoritative assessment of this area of inquiry at the present time. On the other hand, an exhaustive review of this rapidly expanding field would serve to confuse rather than clarify the advances that have been made. Consequently we have decided to emphasize the growing points of the subject and the approaches which promise to advance our knowledge of the chromosome as a physico-chemical entity. Our intention is thus to offer a survey of problems rather than an assembly of facts. It is our hope that this approach will serve as a useful, indeed necessary, introduction not only to this volume of the handbook as such but to all those engaged in the study of the varied and various states of the chromosome as it moves through its ceaseless cycles of synthesis and separation, sexuality and senescence.

I. Cell Cycle and Chromosome Cycle

Asexually reproducing unicells are perpetuated by repeated cell division. They represent the simplest conceivable cell cycle—one which is characterized by a clearly marked sequence of cell division followed by compensatory replication and growth involving both the nucleus and the cytoplasm. This simple cycle is well shown in yeast, though the behaviour of synchronized yeast cultures displaying unbalanced growth shows that the attainment of a particular size is not a prerequisite for division. Rather it is triggered by a mechanism which, in a given environment, operates at regular intervals more or less independent of the size or growth rate of the cell (WILLIAMSON 1966).

In fact natural phasing of cell division via an endogenous diurnal rhythm is not uncommon in unicells and is found, for example, in Dinoflagellates and in the Chlorophyceae. In sexual cycles, on the other hand, one may distinguish four categories of proliferating cell populations:

  (i) Cleaving populations in early embryos and persistent meristems in plants,
  (ii) Differentiating populations in embryonic tissues,
  (iii) Renewal populations such as those in skin, intestinal epithelia and the cork cambium of plants, and
  (iv) Germ line populations.

Of course, in any one organism, cellular events, and hence changes in the character of cell populations, occur in a characteristic temporal pattern. And the information which regulates this pattern, like that which controls spatial organization, must be encoded in the chromosome system.

For diploid plants a simple relationship exists between the minimum mitotic cycle time, the interphase nuclear volume and the DNA content per cell (VAN'T HOF and SPARROW 1963). Thus the cycle time is related linearly to both DNA content per cell and interphase nuclear volume (Fig. 1). Moreover these relationships are independent of chromosome number and the amount of DNA per chromosome (Table 1).

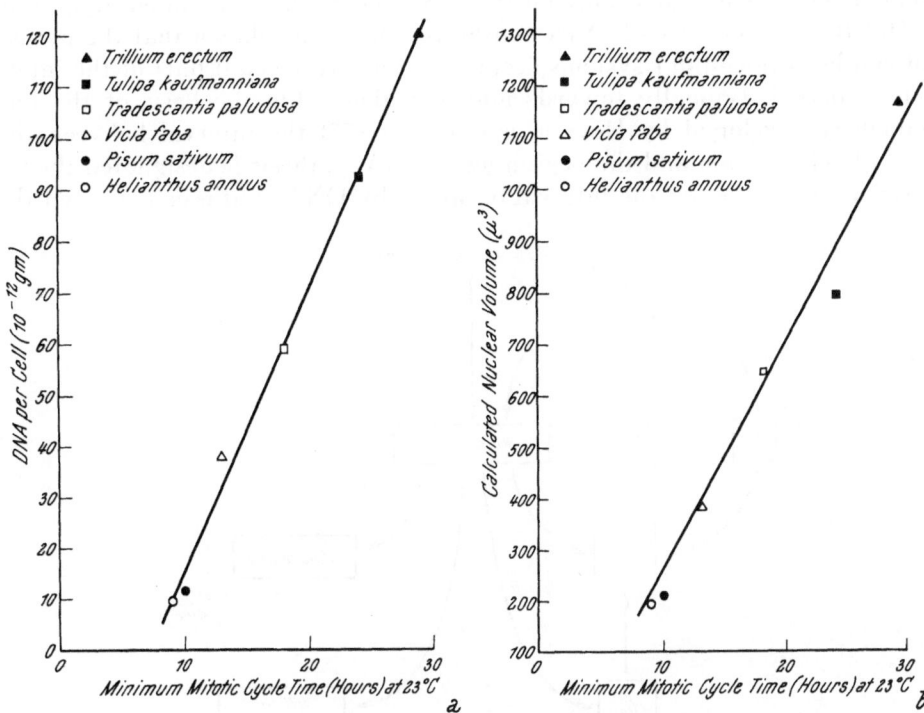

Fig. 1. The relationship between minimum mitotic cycle time and DNA content per cell (*a*) on the one hand and interphase nuclear volume (*b*) on the other. (After VAN'T HOF and SPARROW 1963.)

Table 1. *A Comparison of Nuclear Characteristics of Meristematic Cells from Various Plant Roots.*

(After VAN'T HOF and SPARROW 1963.)

Species	Minimum Cycle Time (hrs.)	Interphase Nuclear Volume (μ^3)	DNA per Cell (10^{-12} gm)	2 n	DNA per Chromosome (10^{-12} gm)
1. *Trillium erectum*	29	1,175	120	10	12
2. *Tulipa kaufmanniana*	23	800	93.7	24	3.91
3. *Tradescantia paludosa*	18	640	59.4	12	4.95
4. *Vicia faba*	13	377	39.4	12	3.2
5. *Pisum sativum*	10	200	11.67	14	0.83
6. *Helianthus annuus*	9	195	9.85	34	0.29

In the steady state renewal system referred to above (type iii) each member is characterized by a cell cycle consisting of two main phases—a synthetic or inter-phase and a mitotic phase. The synthetic phase is concerned with duplicating the essential components of the cell prior to division. As far as the nucleus is concerned this synthesis is essentially an auto-synthesis involving the replication of DNA and, subsequently, the duplication of the chromosome complement. Using the uptake of tritiated thymidine (^{3}H-TdR) as a marker of DNA synthesis it has been shown that the period of synthesis in cellular systems occupies a relatively small part of the interphase period, typically towards mid-interphase. On the basis of the terminology developed by Howard and Pelc (1953) the interval between the end of cell division and the beginning of DNA synthesis is designated the G_1 period (pre-synthetic), the interval occupied by DNA synthesis is termed the

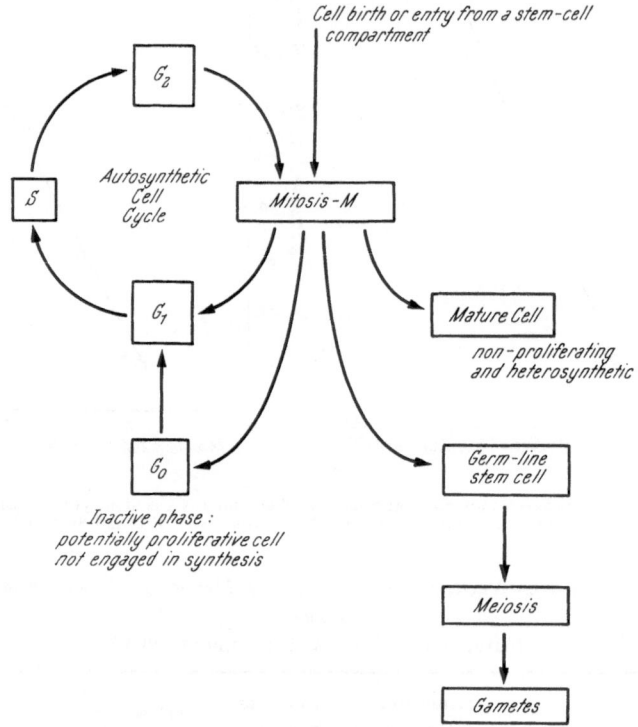

Fig. 2. Components of proliferating cell systems. (Based on Quastler 1963.)

S phase while a further gap phase, G_2 (post-synthetic), occurs between the end of DNA synthesis and the first visible sign of cell division (Fig. 2).

Pure proliferative cell populations are, however, rare. A system usually contains one or several kinds of non-proliferating, mature cells. It may also contain one, or more, groups of potentially proliferative cells which reproduce at a low rate spontaneously or else only upon stimulation.

In a proliferating system there are four possible pathways:

(i) A post-mitotic cell may mature and pass into a differentiated state. Where cellular resources are pre-empted in the building of specialized cell equipment this precludes their alternative utilization for the building of mitotic machinery and the execution of the mitotic act. Thus, certain stable and, possibly, irreversible changes may follow differentiation which result in mitotic suppression or prohibition. Alternatively, cells may become enucleated, so that they can no longer divide (see Section 5 a), or they may become so crowded with specialized components that no space, material or energy is left for a new spindle to form. Of course the mere presence of differentiated structural equipment need not, in itself, prohibit mitosis (see for example p. 71) but it frequently does so (Table 2 and see also p. 72).

(ii) It may serve as a sexual, germ-line, stem cell in which case it will eventually enter a meiotic cycle.

(iii) A post-mitotic cell may complete its cycle and then repeat it by passing into a G_1 phase, or

(iv) It may be shunted into a state of apparent dormancy—a so-called G_0 phase—from which it may subsequently return to a G_1 state. For example, hepatocytes which normally have a very low rate of renewal may remain more-or-less permanently in the G_0 phase. If, as a result of partial

Table 2. *Patterns of Differentiation in Mammalian Cell Types.*
(After BUCHER 1956.)

Cell Type	Capacity for Mitosis	Regenerative Power
1. Renewal tissues, e.g. haemato- poietic systems (spleen, bone marrow) and certain epithelia (intestine, uterus)	Mitotic division usual	Good
2. Liver and renal epithelia	Retained but conserved	Variable
3. Muscle and nerve cells	Greatly reduced or lost	Poor or absent
4. Erythrocytes	No nuclei	Absent

hepatectomy, they are stimulated to divide they then enter into the active cycle by passing into G_1. There is evidence for an inactive G_0 state in other renewal systems too. Thus, WOLFSBERG (1964) has represented the proliferative cycle of the forestomach epithelium of the mouse in the following terms: $M = 1.2$ hrs., $G_1 = 14$ hrs., $S = 13.5$ hrs., $G_2 = 1$–2 hrs., and $G_0 = 360$–704 hrs.

Again, polyploid cells have been reported to divide without a preceding DNA synthesis and in these cases the cells presumably pass directly from G_0 to G_2. Likewise the nuclei of the embryo cells of dry seeds frequently have a DNA value of $4\,C$ (AVANZI et al. 1963, STEIN and QUASTLER 1963). When the seeds germinate the embryo cells pass directly from G_0 to G_2 and to this extent mitosis precedes DNA synthesis.

In multicellular organisms, both plant and animal, the typical auto-synthetic cell cycle extends for approximately 20–24 hours of which the G_1 and S phases respectively occupy about $1/4$ to $1/3$ and $1/2$ of the total time while the G_2 and M periods share the remainder of the time (Table 3). The microspore divisions in plants, however, represent extreme variants. Thus in the first microspore divisions of *Tradescantia* the interphase lasts a week or more. Most of this time is spent in G_1, the S period occurring some 24–36 hours before metaphase (MOSES and TAYLOR 1955). On the other hand the DNA synthesis which precedes the second pollen grain mitosis may be initiated almost immediately following the first, so that the G_1 phase is absent or nearly so (see TAYLOR and McMASTER 1954).

Table 3. *Temporal Components of Cell Cycles.*

| Cell Type | Cycle Time | | | | | Reference |
	M	G_1	S	G_2	Total	
(A) Plant Root Meristems						
1. *Bellavalia romana*	2.75	6	7	6.25	21 hrs.	Jona 1966
2. *Haplopappus gracilis*	1.61	3.46	4	1.44	10.5 hrs.	Sparvoli, Gay, and Kaufmann 1966
3. *Scilla campanulata*	2.3	7.3	10.8	12.2	32.6 hrs.	Evans, G. M., unpublished
4. *Spiranthes sinensis*	3	4	9	6	22 hrs.	Tanaka 1965
5. *Tradescantia paludosa*	3	1	10.5	2.5	17 hrs.	Wimber and Quastler 1963
6. *Vicia faba*	2	4.9	7.5	4.9	19.3 hrs.	Evans and Scott 1964
7. *Zea mays*						
(a) Cap initials	2	– 1	8	5	14 hrs.	
(b) Quiescent centre	3	151	9	11	174 hrs.	Clowes 1965*
(c) Stele just above (b)	2	2	11	7	22 hrs.	
(d) Stele 200 μ from (b)	4	4	9	6	23 hrs.	
(B) Animal Cells in Culture						
1. Chinese hamster fibroblast	0.4	2.7	5.8	2.1	11.0 hrs.	Hsu et al. 1962
2. Mouse fibroblast L-P 59	0.7	9.1	9.9	2.3	22 hrs.	Dewey and Humphrey 1962
3. HeLa P_7	0.5	24	7.0	3.5	35 hrs.	Toliver and Simon 1967
4. *Tetrahymena pyriformis* HSM-strain						
(a) Proteose-peptone medium	30'	50.8'	38.9'	105.3'	225'	Cameron and Nachtwey 1967
(b) Enriched defined medium	25'	99.5'	60.2'	40.3'	225'	
5. *Paramoecium caudatum*-micronucleus	0.5–1	3	3–3.5	0.5–1	8 hrs.	Rao and Prescott 1967

* N. B. See also Table 30.

For a given cell type there is a characteristic cell cycle. For different cells the G_1 phase appears to be the most variable phase. It has been claimed that in adult mammalian tissues the S period has an approximately constant duration of 7 hours. BULLOUGH and LAURENCE (1966) have, however, drawn attention to the fact that this conclusion is based on tissues with relatively high mitotic rates. In tissues where the rate is lower, S may occupy as much as 13.5 hours (mouse forestomach) or even 30 hours (mouse ear epidermis). These authors therefore suggest that the duration of S, like the duration of mitosis, may vary inversely with the mitotic rate. If this is true, then the value of 7 hours may represent an approach to the minimum time required for S in such tissues.

Table 4. *Relationship between DNA Content and Cell Cycle in Angiosperms.*
(Data of VAN'T HOF 1965.)

Species	DNA/cell $(10^{-12}$ gm)	Duration of the Mitotic Cycle (hrs.)	Duration of S (hrs.)
1. *Crepis capillaris*	3.82	10.75	3.25
2. *Impatiens balsamina*	5.14	8.8	3.9
3. *Lycopersicon esculentum*	8.44	10.6	4.3
4. *Allium fistulosum*	41.0	18.8	10.3
5. *Allium cepa*	54.3	17.4	10.9
6. *Tradescantia paludosa*	59.4	20.0	10.8
7. *Allium tuberosum*	66.3	20.6	11.8

When different cell lines from various species of mammals are compared there is a clear tendency for those with the greatest amount of DNA to have an extended S phase. In plants too the duration of S, like that of the cell cycle, increases linearly with the DNA value of the species (Table 4). On the other hand, endoderm cells from neurulae of haploid *Xenopus laevis*, which are half the volume and half the mass of their diploid counterparts, have an S phase which occupies the same relative position in the cell cycle and lasts for the same period as that in diploid cells of the same type (GRAHAM 1966). Here then cell mass cannot be the only factor initiating DNA synthesis for the duration of S is not changed by halving the number of chromosome sets and is not affected by the amount of DNA in the nucleus.

In view of such variations it would seem that the only necessary relationship between the formal phases of the autosynthetic cycle is that chromosome replication must precede cell division. However, there have even been some claims that chromosomes may, on rare occasions, enter mitosis in an unreproduced state and without prior replication (see MAZIA 1961 and ÖSTERGEN 1966). Certainly other components of the standard cycle can be dispensed with. This is especially clear in bacteria where no G_1 or G_2 periods occur. Significantly the "chromosome" here is nothing but DNA and the separation of the replicated DNA is not accomplished by a conventional mitotic process.

Again the myxomycete *Physarium polycephalum* forms multinuclear plasmodia which exhibit naturally synchronous mitoses every 8–10 hours at 26° C. The S period lasts approximately 3 hours immediately following telophase, while G_2 occupies some 6 hours and mitosis itself takes about 1 hour (SACHSENMAIER 1966). There is no G_1 (see Fig. 6). Indeed, even in animals there are cell types in which the major part of interphase is used for DNA synthesis. They belong to rapidly proliferating tissues such as young tissue cultures (WALKER and YATES 1952), the rapidly dividing neuroblasts of locust embryos (BERGERARD 1955, LEACH 1964) and the root tip meristems of plants with a zone of rapid proliferation (WOODARD et al. 1961). Again the cell cycle of embryos may lack one or several of the phases of the standard cell cycle. Thus in *Xenopus laevis* there is no G_1 and G_2 is either very short or else absent during cleavage (Table 5). In the late blastula, how-

Table 5. *Duration, in hours, of Cell Cycle Phases in Developing Xenopus laevis Embryos.* (Data of GRAHAM and MORGAN 1966.)

Developmental Stage	Cell-cycle Phase				Total Cell Cycle (hrs.)
	G_1	S	G_2	M	
1. Mid-cleavage	0	0.17	0.04	0.04	0.25
2. Late-gastrula	2.0	2.0	3.5	0.5	ca. 8
3. Early-gastrula	3.5	4.5	8.0	0.5	ca. 16

ever, there is a substantial G_2 and the G_1 becomes apparent too. From the late blastula to the early tail bud stage the durations of G_1, S and G_2 increase, though the proportion of the whole cycle spent in these phases remains roughly the same. The duration of M, however, remains relatively constant (GRAHAM and MORGAN 1966). This is correlated with a decrease in the rate of cell division during development (compare with DETTLAFF 1964).

Although chromosome duplication must normally precede cell division it need not lead to it. Moreover the total dissociability of the components of the cell cycle can lead to considerable variation on the mitotic theme. Thus cells may become polyploid by virtue of chromosome reproduction without ensuing mitosis and chromosomes may become polytene by virtue of replication without separation of the products of replication (see p. 30). Cells must therefore have distinct regulatory mechanisms which can associate or dissociate the component parts of the cell cycle.

II. Chemistry of the Nucleus

Since DNA forms the essential genetic material it is the prime component of the chromosomes. As a rule the DNA content of the interphase nucleus lies between 10 and 30% of its dry weight (DOUNCE 1952, POLLISTER 1952, ALLFREY et al. 1952). The other substances found in the nucleus vary in amount both in different types of cell and in a particular cell type under different physiological conditions (Table 6). Different nuclei thus differ from each other in their RNA and protein content, their enzymatic equip-

ment and their metabolic activity. They differ also in the distribution of
their chromatin *, the size of their nucleoli and in the presence and pattern
of heterochromatic material. In some nuclei there is so much non-genetic
material that the actual chromosomes represent only a small fraction of the
nuclear constituents whereas compact nuclei are known which contain little
more than genetic material (see p. 88).

One of the main factors underlying the differences in appearance and
stainability of nuclei is variability with respect to protein composition.
Three types of protein are associated with DNA in the interphase nucleus:

(i) Histones—these have long been recognized as the companion protein
of DNA in both plant and animal cells for they are attached to DNA by

Table 6. *Chemical Composition of the Nucleus.*
(After SERRA 1947.)

Component		Quantity	
		μμg per Diploid Nucleus (1 μμg $= 10^{-12}$ gm)	Percent Dry Weight of Nucleus
1. Nucleic Acids	(a) DNA	6–60	5–30
	(b) RNA	0.3–1.5	
2. Proteins	(a) Total protein	24–100	50–80
	(b) Basic protein	10–40	20–40% of total protein
	(c) Non-basic protein	10–60	60–80% of total protein
3. Lipids	(a) Total lipid	5–80	8–40
	(b) Phospholipid	4–70	90% of total lipid

salt-like linkages. Apart from their combination with DNA, the distinguishing features of histones are their basicity and simplicity. They have
molecular weights ranging from 14,000 to 20,000 (BRUNISH et al. 1951). If
there are specific somatic tissue differences between histones they have
not yet been demonstrated though some 10–15 types of histone are now
known (BUSCH et al. 1964). Almost nothing is known about the structural
features of histones. Both ASTBURY (1947) and WILKINS and RANDALL (1953)
have found that the X-ray diffraction pattern of nucleoprotein is very
similar to that of DNA alone. This suggests that the protein component,
or at least some of it, also assumes a helical form. It is often claimed that
histone, like DNA, is virtually confined to the cell nucleus. But there is
evidence that histones can be synthesized in the cytoplasm (see p. 76).
Indeed, LESLIE (1961) claims that histone-like proteins are present in ribosomes prepared from plant and animal cells. While ribosomal protein is
characterized by high contents of lysine and arginine, in addition to aspartic

---

* The native complex of DNA with histone and non-histone protein, as well
as with small amounts of RNA is commonly referred to as "chromatin".

Table 7. The Classification of Nucleic Acid Types in HeLa Cells.
(After Warner 1967.)

N. A. Species		Origin	Function	Base Composition			
				G	C	A	U (T)
Nuclear	DNA	—	—	22	21	29	28
	45S-RNA	De novo	Precursor to r-RNA species	37	28	15	19
	32S-RNA	45S-RNA	Precursor to 28S-RNA	36	33	15	16
Cytoplasmic	28S-RNA	32S-RNA	Major part of large ribosomal sub-unit	36	32	16	16
	16S-RNA	45S-RNA	Major part of small ribosomal sub-unit	30	27	21	22
	5S-RNA	32S-RNA	Part of large ribosomal sub-unit				
	m-RNA	DNA	Messenger RNA	21.5	25.5	26	27
	t-RNA (= 4S-RNA)		Transfer RNA	30	25	21	24

and glutamic acid, it is distinguishable from histone by its lower content of basic amino acids, by the presence of tryptophan and by its relative insolubility in acid (BONNER 1965).

(ii) Acidic Proteins—these are a heterogeneous group of proteins with, however, similar solubilities, amino-acid composition and NH_2-terminal amino-acids. These proteins all contain 8—10% of aspartic and 11—14% of glutamic acid as well as substantial amounts of leucine and glycine (BUSCH et al. 1964). They are present in the nucleolus, the nuclear sap and the chromatin and their amino acid composition closely resembles that of the ribosomal proteins. Their turnover rate is considerably greater than that of the histones and in smaller nuclei there is much less acidic protein in the nuclear sap.

(iii) Enzymic Proteins—these include the enzymes which are of importance in energy-yielding reactions necessary for nuclear metabolism as well as the metabolic enzymes themselves (MIRSKY and OSAWA 1961).

As far as nuclear RNA is concerned two principal classes can be distinguished:

(i) Two fractions which sediment in the centrifuge at 45 Svedbergs (S units) and 32 S respectively. The base composition of both these RNA species closely resembles that of ribosomal (r) RNA (Table 7). This suggests a precursor-product relationship which is confirmed by blocking RNA synthesis. Thus if actinomycin-D is added to cells which have been incorporating ^{14}C-UdR for 20 minutes and the radioactivity of successive sucrose gradients is followed, there is a shift of radioactivity from 45 S to 32 S and thence to 28 S and 16 S (WARNER 1967). The 28 S and 16 S particles, which make up the major portion of the ribosomal subunits (which sediment at 50 S and 30 S respectively) are, therefore, formed in the absence of RNA-synthesis provided 45 S RNA is present.

(ii) Heterogeneous (HS) RNA. This has a base composition which strongly resembles that of the cell's DNA. Neither the fate nor the function of HS-RNA has been satisfactorily explained. The only reasonable suggestion is that it represents a precursor to m-RNA. Both these categories of RNA share a number of distinctive features including: (a) The DNA-like base ratio, (b) Heterogeneity of size, and (c) The apparent lack of secondary structure.

III. Metabolic Nuclear Activity

From a functional point of view the chromosome system of plants and animals performs three primary roles. First, it is a replicator during the duplication of the genetic information. Secondly, it is a transcriber by which this genetic information can be made available for the control of the metabolic activity of the cell. And thirdly, it is a repository for conserving genetic information. Indeed much of the complexity of the chromosome cycle would appear to reflect the necessity for the efficient handling of the large amount of DNA present in the system.

From a synthetic point of view it is necessary, therefore, to distinguish between three principal states of chromosome activity—the autosynthetic, the heterosynthetic and the non-synthetic. The first characterizes cells which are engaged in active proliferation and it prepares them for a series of successive mitoses. The second is found in those cells which are differentiating and are involved in the production of novel, specific molecules. The last occurs in cells which have completed differentiation and either show no activity at all or else are concerned with maintaining a limited and specific pattern of metabolic activity.

In order to appreciate the distinction between these three states, and the varying patterns of chromosome organization which determine them, we need to remind ourselves of the relationship between the flow of genetic information in biological systems and the regulation of synthesis. All molecular activities in living systems depend ultimately on the transmission of chemical information. This information is sequestered in the linear sequence of complementary base pairs, coded in triplets, which occur in the DNA dimer. The deciphering of this information depends upon a form of molecular symbiosis between different nucleic acid species. This, in turn, specifies the amino acid sequences and shapes assumed during protein synthesis. The process thus involves two main steps. First, the transcription of the information in DNA onto a messenger RNA molecule (m-RNA). Second, the attachment of m-RNA to a ribosome sequence and the translation of the transcribed base sequence into a specific amino-acid (AA) sequence. This latter event requires the mediation of a series of transfer RNA molecules (t-RNA) which, in the presence of appropriate activating enzymes, will complex with specific AA molecules. DNA is thus a master template from which secondary working templates may be constructed.

The regulation of this scheme of protein synthesis is accomplished, at least in part, by systems of negative feedback operating either:

(i) By end-product inhibition whereby the catalytic activity of an enzyme operative early in a synthetic sequence is blocked when the concentration of the final product of that sequence reaches a particular level, or

(ii) By repression whereby the accumulation of a synthetic product blocks the primary or secondary templates responsible for the production of the enzymes necessary for its synthesis.

Repression is clearly a more efficient type of feedback control than inhibition since the activities of the cell are not wasted in producing unwanted enzymes. But it is slower in its action than inhibition which, even if of limited economy, is both rapid and highly flexible.

In bacteria and viruses there is now good evidence that repression is governed by specific regulator genes, which control the amount of enzyme produced, as distinct from structural genes, which control its AA composition and, thereby, its specific structure and function. There is also evidence that a third species of gene—the operator—is involved. This operator is contiguous with one or more structural genes which are transcribed as a unit. The operator gene controls the initiation of local m-RNA synthesis. It is assumed that the regulator gene produces a protein which can hybridize with the operator and so repress its action (Fig. 3).

There has been a general tendency to assume that equivalent regulatory genes occur also in plants and animals though, to our knowledge, they have not yet been demonstrated. There is, however, a growing body of evidence which points to the existence of a distinct system of regulation which is not apparent in bacteria or viruses. This system depends on the condensation of chromosomes or chromosome regions and leads to repression of entire blocks of genes, rather than micro-regions of the genotype (see p. 65).

Now in autosynthesis DNA functions as a template in its own enzyme-mediated replication. In heterosynthesis an enzyme transcription of a nucleic acid template occurs but the transcribed product is then itself translated into a sequence of AA's by a further set of processes. At the root of the distinction between these two forms of synthesis lies the central dogma of translation. This states that while a polypeptide chain can be derived from a polynucleotide sequence the reciprocal action is impossible; nor can specificity be transferred from one polypeptide to another. In essence the distinction between auto- and heterosynthesis is between nuclear metabolism which serves the chromosome and that which serves the cytoplasm. Both kinds of synthesis, however, depend on the state of the chromosome material and the diffuse state which characterizes the chromatin of the interphase nucleus serves both for replication and for transcription.

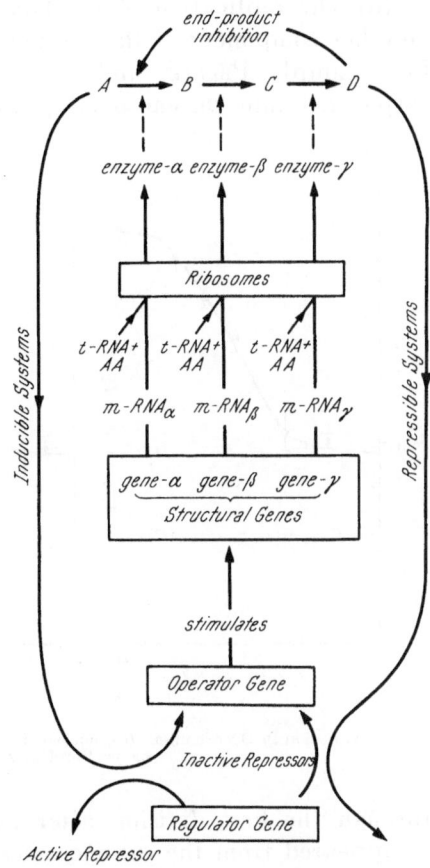

Fig. 3. Systems of gene regulation.

1. Chromosome Reproduction

a) Mitotic Cells

Essentially the autosynthetic phase is one of chromosome duplication. In molecular terms this means the production of all those molecules which give the chromosome form, regulate its metabolism and control its mechanical activity during the cell cycle. The primary event in this process is undoubtedly the replication of DNA. Two lines of argument support this conclusion:

(i) DNA is the only self-replicating molecule known in the nucleus and, in consequence, the synthesis of any other chromosome constituent must, directly or indirectly, depend on heterocatalytic activity. That is, they must be synthesized through the derivative action of DNA.

(ii) The replication of the DNA molecule is semi-conservative whereas no other component of the chromosome has yet been found to be conserved. For example, Prensky and Smith (1964) have shown that ³H-arginine incorporated into chromosomal proteins remains a part of the chromosome

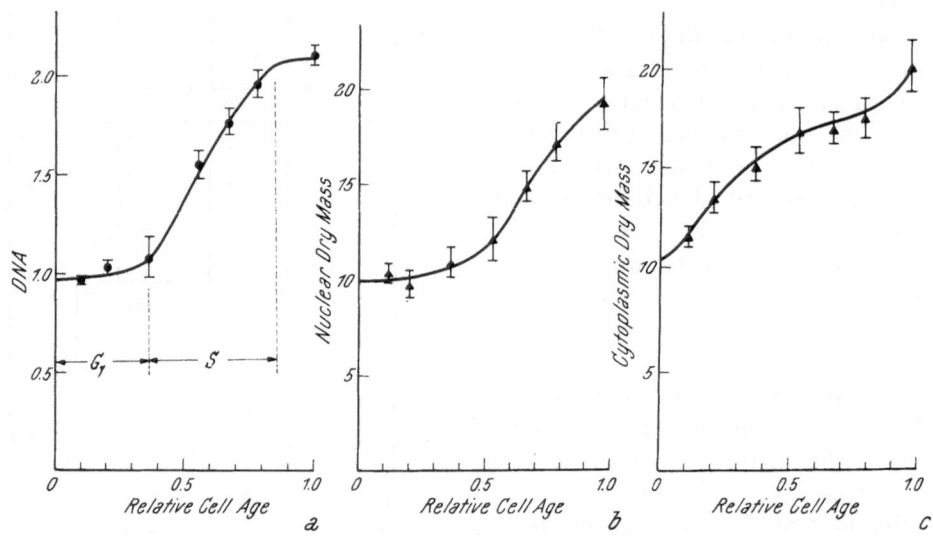

Fig. 4. Variation in DNA content (*a*), nuclear dry mass (*b*) and cytoplasmic dry mass (*c*) during interphase in mouse fibroblasts. (After Zetterberg 1966.)

through the first division after incorporation but has almost completely disappeared from the chromosomes after one replication. In this case most of the arginine was presumably incorporated into histone.

The optional G_1 and G_2 phases of the autosynthetic cycle represent periods to which, as yet, it has not been possible to ascribe with certainty any specific biochemical events of the cell cycle. There is certainly evidence for both RNA and protein synthesis during these phases, as well as during S, but the significance of this synthesis in advancing the mitotic cycle remains debatable. It has been argued that a part of the nuclear proteins produced in G_1 and S is necessary for the initiation and maintenance of DNA replication whereas proteins produced in G_2 and prophase may be concerned with chromosome coiling. Shapiro and Levina (1967), using pulse labelling with ³H-arginine and ³H-lysine, found that nuclear proteins containing arginine and lysine begin to form at G_1. Their synthesis is continued into S and G_2 but at a much reduced rate. Zetterberg (1966) has presented interferometric data on mouse fibroblasts *in vitro* which show that the patterns of increase of mass in cytoplasm and nucleus during interphase are very different (Fig. 4). Thus, whereas during G_1 the nuclear mass remains constant, that of the cytoplasm increases markedly. The major increase in nuclear mass takes place when DNA synthesis is initiated. Coincident with this, the rate of mass accumulation in the cytoplasm decreases.

It would appear, therefore, that during G_1 most of the protein being synthesized in the cell is accumulated in the cytoplasm. The rate of protein synthesis per unit cytoplasmic mass is, however, constant throughout inter-

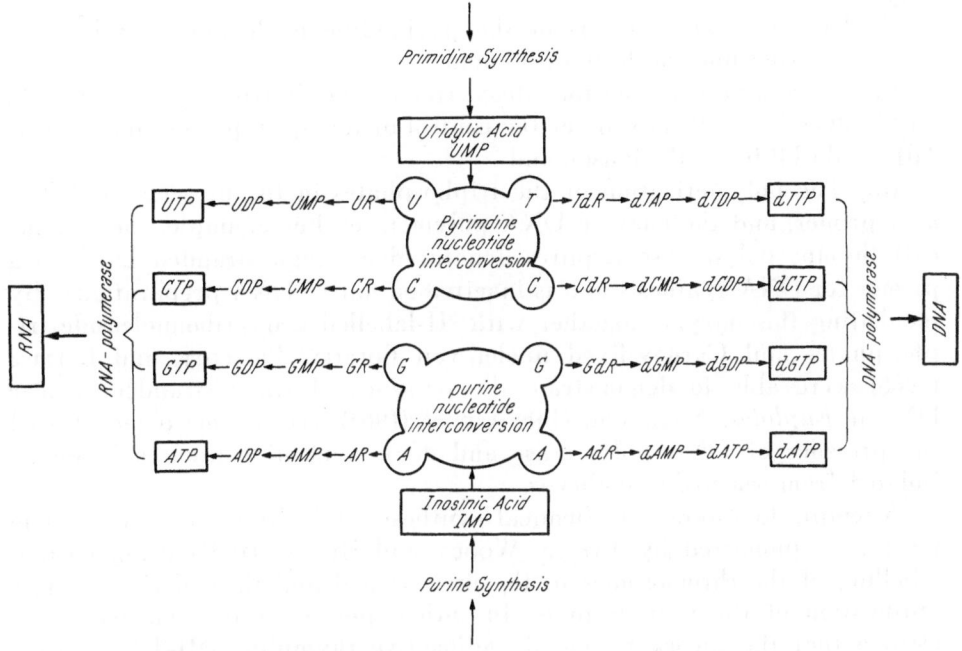

Fig. 5. The principal phases of nucleic acid synthesis. (After KIHLMAN 1967.)

phase. Since less protein accumulates in the cytoplasm during S this suggests that there is a movement of protein from the cytoplasm to the nucleus. And ZETTERBERG believes that the proteins involved in this movement are non-chromosomal ones.

A further indication that the induction of S is dependent on material from the cytoplasm comes from the experiments of GRAHAM et al. (1966). They have shown that, while nuclei from certain fully differentiated adult cells of Xenopus laevis do not enter mitosis after transplantation into egg cytoplasm, many of these nuclei are induced to synthesize DNA. These include nuclei from liver, brain and blood cells, less than 1% of which normally synthesize DNA in any two-hour period. Clearly, in this instance, the egg cytoplasm is capable of inducing all nuclei with unreplicated chromosomes to commence DNA synthesis. That a particular protein must be made which controls the initiation of replication is suggested also by the work of E. W. TAYLOR (1965). Thus, he succeeded in reducing DNA synthesis to 10—20% its normal value in cultures of human cells (strain KB) by the addition of puromycin which is known to inhibit protein synthesis. Assays of DNA polymerase and thymidine kinase indicated that synthesis of these enzymes after thymidine addition could not account for the requirement for protein synthesis. Finally, Terasima and Yasukawa (1966) also argue that a certain amount of protein synthesis during G_1 is necessary for the commencement of DNA synthesis and that this G_1 protein is different from the proteins synthesized in association with DNA replication itself.

The process of DNA synthesis itself involves three distinct phases (Fig. 5):

(i) The *de novo* synthesis of the pyrimidine nucleotide uridylic acid and the purine nucleotide inosinic acid,

(ii) The synthesis of the four deoxyribonucleoside triphosphates (dATP, dTTP, dGTP and dCTP) under the control of the appropriate kinases (e.g., TdR → dTTP by TdR-kinase), and

(iii) The polymerization of the triphosphates in the presence of DNA, as a primer, and the enzyme DNA polymerase. For example, the enzyme calf thymus polymerase requires denatured or single-stranded DNA as a primer for DNA synthesis and is inactive on native DNA preparations. By incubating this enzyme together with ^{3}H-labelled deoxyribonucleoside triphosphates and Carnoy-fixed nuclei von Borstel, Prescott, and Bollum (1966) were able to demonstrate the presence of single-stranded primer DNA in *Euplotes*. Mazia and Hinegardner (1963) have in fact demonstrated the presence of DNA polymerase and deoxynucleotide kinases in nuclei isolated from sea urchin embryos.

Attempts to follow the chemical synthesis of DNA at the chromosome level were pioneered by Taylor, Woods, and Hughes (1957) using tritium labelling of the chromosomes in the first, second and third divisions after application of the radio-isotope. In such experiments it is customary to assume that the incorporation of radioactive thymidine (^{3}H-TdR) into a chromosome implies the occurrence of replication. The possibility that the uptake of the nucleotide by the chromosome need not imply its incorporation into a growing polymer has been given scant consideration presumably because the limits of S are so well defined. Moreover, thymidine kinase formation can be detected only at the beginning of the S phase in the mitotic cycle (Fig. 6).

In Taylor's original experiment isotope was available in the culture medium in which roots of *Vicia faba* were grown for a time period approximately equal to the duration of one replication period. After this the material was transferred to colchicine solution. This, in addition to increasing the frequency of metaphases available for examination and facilitating the separation of both chromosomes and chromatids, permitted the number of post-treatment replication cycles to be followed. Thus, cells without any further duplication after transfer to colchicine had a normal diploid chromosome number ($2n = 2x = 12$). Cells where the chromosomes had undergone one post-treatment duplication contained 24 ($2n = 4x$) chromosomes, while those in which two such events had occurred had 48 ($2n = 8x$).

In cells with 12 chromosomes (X₁) all were reported to be uniformly labelled over both sister chromatids. In the endo-tetraploid X₂, however, while all chromosomes were labelled, the label was usually confined to one of the two sister chromatids in each chromosome. The only exception to this pattern of all-or-none segregation at the X₂ stage was the occasional com-

plementary distribution of label over both chromatids such that homologous regions of sister chromatids showed label segregation. In these cases it was postulated that an exchange had occurred between sister chromatids. Finally in 8 x (X₃) cells one-half of the complement had chromosomes with one labelled and one unlabelled chromatid while the other half was without label.

The same general pattern of isotope distribution (Fig. 7) has subsequently been reported in *Bellavalia* (TAYLOR 1958), *Crepis* (TAYLOR 1958), *Tradescantia* (WIMBER unpublished, see PEACOCK 1965), *Allium* (NORDQVIST unpublished, see LIMA-DE-FARIA 1962), the Chinese hamster (TAYLOR 1960, PRESCOTT and BENDER 1963) and *Potorous* (WALEN 1963). These experiments strongly support a semiconservative scheme of chromosome replication. Such a scheme implies that the chromatid must be composed of complementary subunits which remain intact (except, of course, at points of sister-chromatid exchange) during replication. It is tempting, therefore, to think of the chromatid as nothing more than a twin DNA helix and many have succumbed to this temptation.

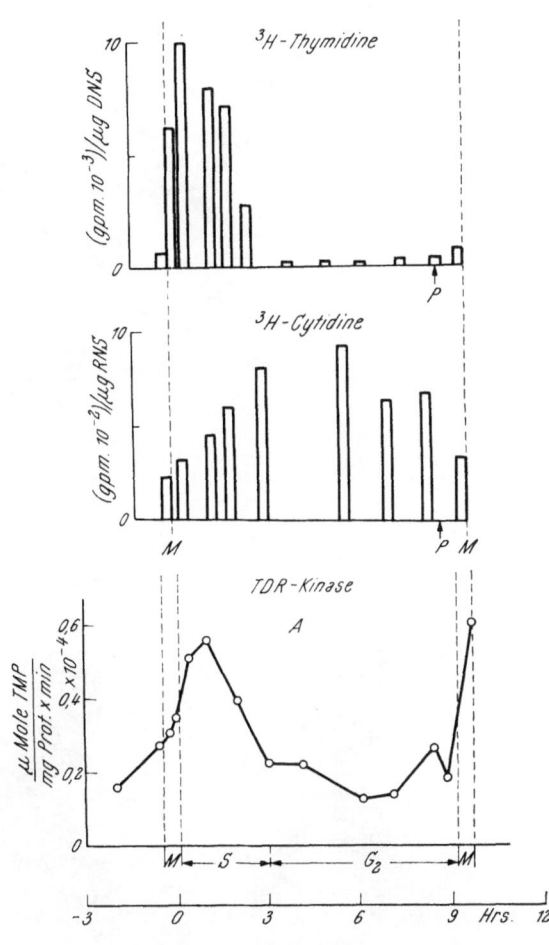

Fig. 6. The synthesis of DNA, RNA and thymidine kinase during the cell cycle of *Physarium polycephalum*. M = mitosis, P = prophase. (After SACHSENHEIMER 1966.)

Significant exceptions to the general pattern reported by TAYLOR have, however, been found. The first of these stems from the work of LA COUR and PELC (1958) who repeated the experiment of TAYLOR et al. both with and without colchicine to test whether this chemical had any effect on the replication process. By applying colchicine both before and after exposure to ³H-TdR LA COUR and PELC found some X₁ chromosomes with only one sister-chromatid labelled. When, on the other hand, colchicine was not present at either the first or the second replication most X_2 chromosomes were labelled in both sister chromatids. Indeed, even when the same colchicine treatment as that employed by TAYLOR, WOODS, and HUGHES was used they found some X_2 cells in which both sister chromatids were labelled.

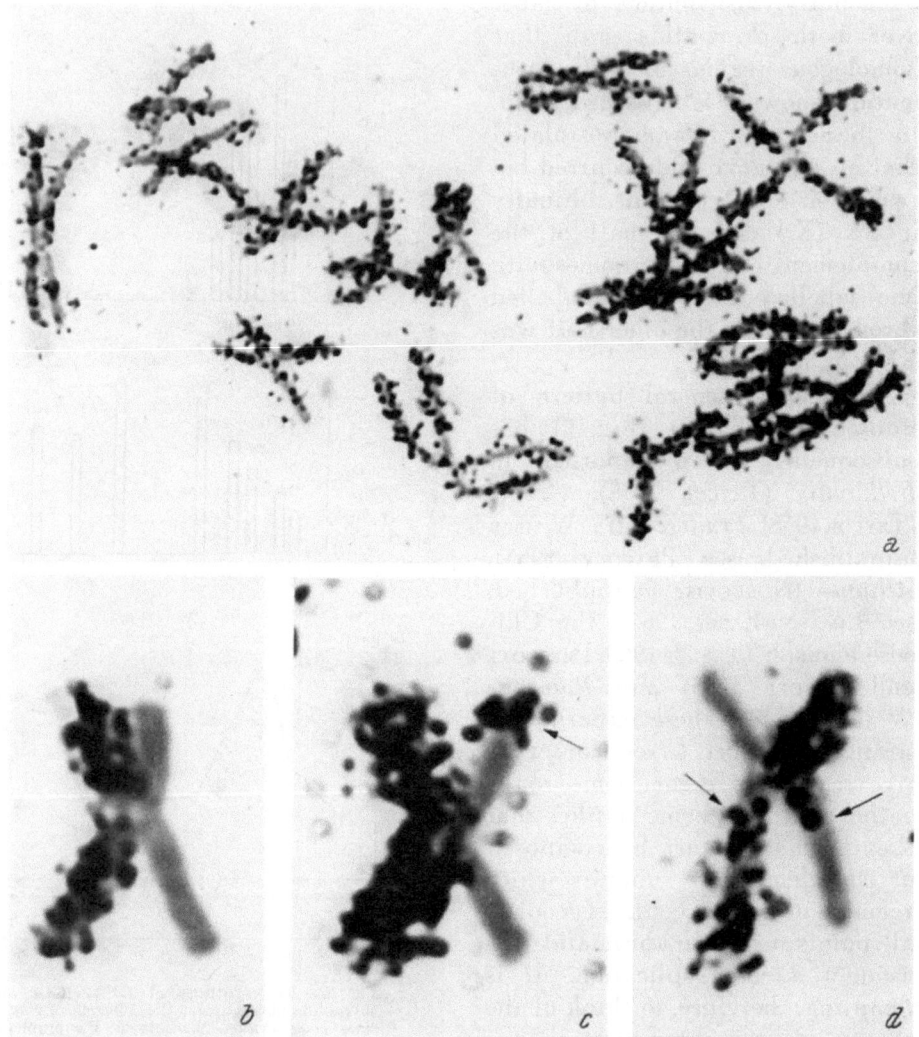

Fig. 7. Distribution of ^{3}H-TdR in X_1 (a) and X_2 (b—d) chromosomes of *Allium*. (Photographs kindly supplied by Dr. D. P. FOX.)

In a study allegedly designed to check LA COUR and PELC's findings, WOODS and SCHAIRER (1959) carried out two experiments in which no colchicine pre-treatment was given prior to isotope application. In both, the concentration of colchicine was appreciably higher than that employed by LA COUR and PELC and the concentration of thymidine was also different. From their results they argued that colchicine had no effect on

DNA replication and suggested that faulty squashing could account for LA COUR and PELC's findings. The latter, however, rightly argue that WOODS and SCHAIRER's results are not sufficiently accurate to detect a change due to colchicine since only 50 chromosomes were studied by them; this, of course, represents only slightly more than the complements of four complete cells.

X_1 results similar to those of LA COUR and PELC have been reported from *Allium cepa* following the same treatment employed by TAYLOR et al. in *Vicia* (NORDQVIST unpublished in LIMA-DE-FARIA 1962) but no other worker seems to have found departures in X_1 behaviour. The issue thus remains unresolved.

The exceptional X_2 results found by LA COUR and PELC have since been partly acknowledged, though not to our knowledge adequately explained, by TAYLOR (1958) himself. He reports that in rare instances he too finds X_2 chromosomes with both chromatids labelled correspondingly for a part of their length. PEACOCK (1965), on the other hand, finds a significant proportion of X_2 chromosomes in *Vicia faba* (241/1164) to show an iso-labelling of sister chromatids. This was sometimes restricted to a region of a chromosome which elsewhere had label segregation but in other cases appeared to extend along the full length. Segmental iso-labelling most frequently proved to be terminal (Fig. 7 c) but iso-labelled segments could also occur in the intercept between regions of label segregation (17/241). Iso-labelling at X_2 has now been reported for other organisms investigated by the technique of tritium autoradiography (see PEACOCK 1965).

TAYLOR has suggested that iso-labelling results from interchromosomal exchange in the colchicine tetraploid cells. On this scheme there ought to be a correspondingly iso-unlabelled region on the other homologue. PEACOCK (1965) claims that at least in some cases this can be rejected on positive grounds. This author has therefore taken iso-labelling to mean that there must be at least two double DNA helices per chromatid. Asynchronous replication between the component molecules in such a multistranded chromosome could, of course, yield label segregation or inequality of label at X_1 and both these conditions, as we have seen, have been claimed.

Despite such complications there seems little doubt that the mechanism of chromosome duplication, like that of DNA replication, is semiconservative, though the steps leading to the initiation of DNA synthesis and the segregation of DNA into chromatids remain obvious gaps in our knowledge of the chromosome cycle. Indeed, until the actual number of DNA molecules per chromatid can be unambiguously resolved the equating of chromosome duplication with DNA replication cannot be regarded as acceptable (see also pp. 91—98).

Since considerable variation may exist in the rate of cell development during the autosynthetic cycle it is not surprising to find that not all cells at any given time show exactly the same pattern of labelling. Neither, however, are all chromosomes in any one nucleus necessarily synchronous in either the initiation or the completion of their replication. Even so each DNA molecule normally replicates only once. Replication must therefore in some way change DNA so that no further replication can occur until the next S phase. That this is not an intrinsic property of DNA is clear from the fact that the replication of viral DNA is not subject to this limitation. Likewise in the diplosporous apomict *Allium odorum* an extra replication takes place regularly on the female side. This compensates for the lack of fertilization (Håkansson and Levan 1957).

There is now considerable evidence from autoradiographic studies involving pulse labelling to suggest that DNA synthesis is initiated at a number of different points within individual chromosomes. Thus in both plants and animals ^{3}H-TdR is incorporated at many separate sites in single chromosomes after pulses which are short compared to the time required for complete DNA replication (see also p. 92). The existence of multiple replication points goes some way towards resolving the problem of how the immense length of the DNA present in the chromosomes of plants and animals is able to carry out its replication so rapidly. Thus, in the Chinese hamster, total DNA synthesis takes about 6 hours. Since there is about 9 cm of DNA in an average Chinese hamster chromosome then with only one replication point per chromosome it would require some 15 hours to replicate all the DNA even at the fast bacterial rate of synthesis of 100 μ per min (Huberman and Riggs 1966).

Recognizable patterns of asynchrony exist not only between, but also within, individual chromosomes. The most striking of these, and the one in terms of which most variants can be interpreted, is that associated with the distribution of eu- and hetero-chromatic regions in the complement. Numerous reports have confirmed the claim that the standard-type heterochromatic X-chromosome found in the mitotic nuclei of many female mammals labels late in the synthetic sequence. For example in a culture of human female fibroblasts, Comings (1967 a) found that the heterochromatic X did not begin DNA synthesis until 2.5 hours after the initiation of DNA synthesis in the euchromatic regions. It then replicated at a rate approximately equal to that of euchromatin for about 3.5 hours and continued synthesis for about 1.5 hours after replication had been completed in the euchromatin. The total time taken for the replication of the heterochromatic X was thus 5.0 hours compared with the 6–7 hours taken by the euchromatic material.

By exposing fibroblasts in culture to ^{3}H-TdR for only 3 min prior to fixation and autoradiographic analysis, Comings (1967 b) also claims to have shown that the sex-chromatin body retains its heteropycnotic form during DNA replication. His argument is as follows: Following flash exposure, grains on the autoradiograph should occur only over areas synthesizing DNA at the time the cells were fixed. If heterochromatic material must uncoil or unfold in order to replicate then no grains should appear over any sex-chromatin bodies. In fact, of 188 labelled fibroblasts examined, 138 had a heteropycnotic sex chromatin body and 91 of these (66%) were clearly covered by grains. Indeed in 29 the density of grains over the body exceeded that over the remaining euchromatin. In other mammalian species where a more complex pattern of female somatic heteropycnocity occurs this too is paralleled by an equivalent late-labelling sequence (OHNO et al. 1964, WOLF et al. 1965, GALTON et al. 1965).

An equivalent disparity of labelling can be identified also in cases where no sex chromosomes are present but where heterochromatin can be recognized. For example, in *Vicia faba* the heterochromatic zones are difficult to distinguish at metaphase but appear as relatively deeper staining segments in pro- and telo-phase nuclei and are clearly observed in interphase as discrete Feulgen-positive blocks. From observations on interphase nuclei LA COUR and PELC (1958) suggested that the hetero- and euchromatic regions in *Vicia* underwent replication at different times. More recently EVANS (1964) has shown that, in X_1 cells which receive isotope during the last 2 hours of S, the label is mainly confined to the heterochromatic zones which occur predominantly on either side of the centromere in the large metacentric M-chromosomes and in the mid regions of the acrocentric S-chromosomes. Here too then synthesis is delayed in heterochromatic segments.

In species, like the mouse, where heterochromatic material is localized at the centric regions, late labelling of these regions could be erroneously construed as indicating that late replication proceeds from the distal to the proximal ends of the chromosome. This in fact has been claimed for *Crepis capillaris* (TAYLOR 1958), a species where pro-centric heterochromatin is known to occur.

The reason for this specific delay in DNA replication at heterochromatic sites is not known*. Certainly the relationship is not an invariable one. In *Spiranthes sinensis,* for example, TANAKA (1965) has described a quite different relationship. Here DNA synthesis occurs earlier in heterochromatic regions.

Yet a further variant is found in the mealy bug, *Planococcus citri.* In males of this species the paternal set becomes positively heteropycnotic in early embryogeny giving rise to 5 H and 5 E chromosomes. Following this the heteropycnotic set becomes late replicating (BAER 1965). Now the cells of the sheath covering the testis are of different size, a consequence of endomitosis. The smallest cells undergoing endomitosis had about 40 E chromosomes and a H-body about the same size as that found in the smallest non-dividing cells of this tissue. The largest sheath cells had about 80 E chromosomes and again the H-body was about the same size as that of the smallest cells. Indeed in some of these cells 5 H chromosomes could be seen in the body. This means that while the E-chromosomes had undergone 3 or 4 cycles of replication, the H-chromosomes had not replicated at all (NUR 1966). Thus in certain polyploid cells the H-set fails to replicate while the E-set undergoes several successive cycles of replication (see also p. 38).

Most of the observations on ³H-TdR labelling which we have so far mentioned are qualitative in character. EVANS and REES (1966) and REES and EVANS (1966) have recently made a quantitative study of tritium uptake in *Scilla campanulata.* This is a species with 8 distinguishable chromosome pairs where the little heterochromatin present is scattered in small segments throughout the complement. Using the first group of labelled nuclei

---

* See note 1 Appendix.

to reach metaphase following a 30 min exposure to ^{3}H-TdR, they report that:

(i) Variation in the amount of DNA replication, as judged by the variation in frequency of silver grains between chromosomes at late S, is directly proportional to chromosome length.

(ii) Within chromosomes, labelling was disproportionately low in distal chromosome segments and in segments proximal to the centromere. These two effects were, however, often confounded because in most chromosomes the centromeres are subterminal. Chromosome V is exceptional in having a median centromere and, in this chromosome, the distribution of silver grains indicates that the centromere and end effects are of the same order of magnitude. These results show that DNA replication is completed earlier near to the centromeres and the ends than elsewhere.

(iii) The pattern of DNA replication at late S of mitosis is correlated with the distribution of chiasmata at meiosis. Chromosome regions where the amount of late DNA replication per unit length of chromosome is highest (interstitial regions) have the highest chiasma frequencies (Fig. 8).

In *Puschkinia libanotica* Barlow (unpub.) also found a relationship between chiasma density and grain density in segments labelled during the last 18 minutes of the S phase. But in this case the correlation was negative.

One final point of interest relates to the role of the nucleolus in controlling DNA synthesis. Das (1962), from a study on micronuclei derived from X-ray induced chromosome fragments in *Allium cepa* and *Vicia faba*, showed that the ability of a micronucleus to synthesize DNA can be correlated with the presence of persistent nucleolar bodies (Table 8). Scott and Evans (1964) obtained comparable results in the case of micronuclei induced in root tip cells of *Vicia faba* following exposure to maleic hydrazide (MH). Significantly, exposure of nucleoli to high doses of microbeam irradiation is known to attenuate, though not abolish, DNA synthesis (Perry, Hell, and Errera 1961).

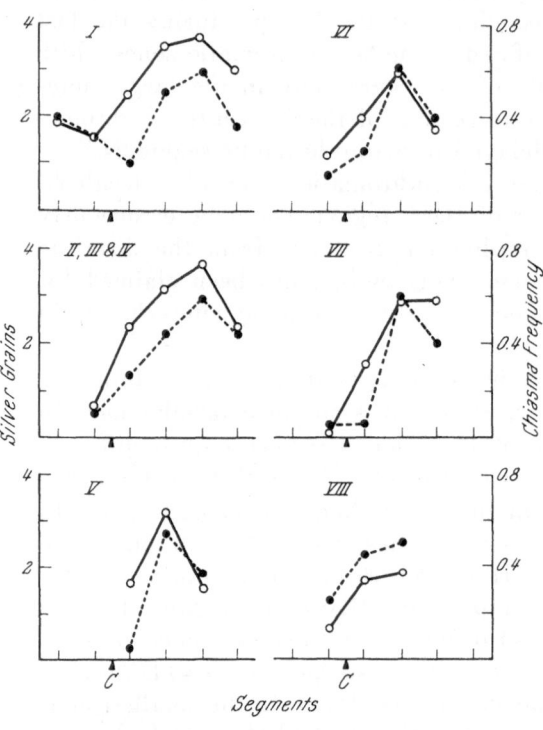

Fig. 8. Distribution of mean chiasma frequencies (- - ○ - -) and mean grain counts (— ○ —) following ^{3}H-TdR labelling at late S in the chromosomes of *Scilla campanulata*. The position of the centromere (c) is indicated by an arrow. In chromosome V, which is metacentric, only one arm is shown. (After Rees and Evans 1966.)

b) Meiotic Cells

There is now unambiguous evidence that the synthesis of DNA in preparation for a meiotic cycle occurs principally during late pre-meiotic interphase or else at leptotene (see Table 71 JOHN and LEWIS 1965). There is also evidence that the segregation of labelled DNA at this synthesis is semiconservative. Thus in the grasshopper *Romalea* the first labelled cells to arrive at anaphase-I and metaphase-II (24 days after ^{3}H-TdR labelling) have both chromatids of each half-bivalent labelled (TAYLOR 1965). By contrast, cells observed at anaphase-I on the 28th post-treatment day showed label segregation. These were considered by TAYLOR to have incorporated ^{3}H-TdR at the second interphase preceding meiotic prophase. Finally, the pre-

Table 8. *Relationship between the Frequency of Labelled and Unlabelled Micronuclei with and without Nucleolar Material.*
(After DAS 1962.)

Species	Exposure to ^{3}H-TdR (hrs.)	Labelled Micronuclei		Unlabelled Micronuclei	
		Total	Percent with Nucleolar Material	Total	Percent without Nucleolar Material
1. Onion	2	5	80.0	65	9.3
	4	19	73.6	88	4.5
2. Bean	2	88	90.9	185	3.8
	4	100	64.0	195	0.5

meiotic DNA replication of heterochromatic sex chromosomes is retarded (Fig. 9) relative to autosomal DNA replication (LIMA-DE-FARIA 1959, 1962 b, HENDERSON 1964).

HOTTA, ITO, and STERN (1966) and ITO, HOTTA, and STERN (1967) have, however, argued for a further and separate period of DNA synthesis during the first meiotic prophase. This argument is based on a study of the biochemical events which take place during the meiosis of liliaceous plants. In these studies, as indeed in others carried out by them both on meiosis and the ensuing pollen grain mitosis (see pp. 48 und 85), they have used the correlation between bud length and developmental stage first established by ERICKSON (1947, 1948). VASIL (1967) has, however, pointed out that anthers and their component tissues are extremely susceptible to even mild changes, whether external or internal in character. This, coupled with the fact that a gradient of stages can be found from the base to the apex of the anther, has led him to question the validity of their technique. He argues that a slight difference, or a mistake, in the determination of the exact stage may lead to incorrect conclusions. VASIL has also claimed that the synchrony of meiotic and post-meiotic stages is lost to a considerable extent under *in vitro* conditions, a claim in direct opposition to that of STERN and his colleagues. Bearing these criticisms in mind let us examine the evidence for a prophase synthesis of DNA at meiosis. This evidence is of two kinds:

(i) By exposing intact flower buds of *Lilium longiflorum* and *Trillium erectum* to ^{32}P-phosphate (20 μ/ml) for 24 hours during different intervals of the meiotic cycle, two phases of labelling were detected. DNA isolated

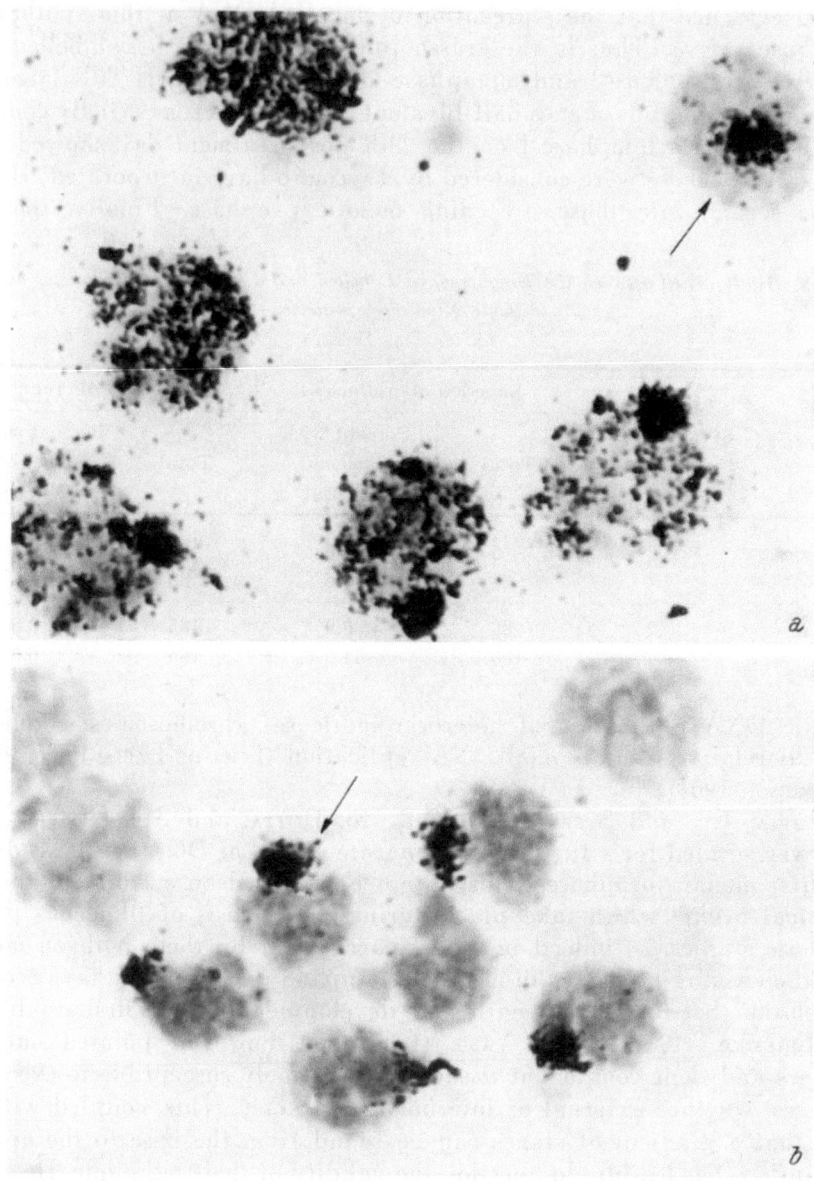

Fig. 9. Delayed DNA synthesis in the heterochromatic X chromosomes of *Tettigonia cantans* (a) and *Achaeta domestica* (b) at first prophase of meiosis. (Photographs kindly supplied by Dr. S. A. Henderson, ca. ×1,500.)

from cells in pre-meiotic S was heavily labelled while that isolated from zygotene-pachytene was lightly labelled. No appreciable label was found in the DNA of cells exposed to isotope between the termination of the

S phase and late leptotene on the one hand or at any or all of the meiotic stages after pachytene on the other. Labelling with ³H-TdR yielded similar results and the pattern was abolished by treatment with DNase though not by RNase or protease treatment. Tests made on isolated nuclei gave the same result so excluding the possibility that the synthesis was cytoplasmic in character.

It is true that the amount of radioactivity incorporated into the DNA during meiotic prophase was very small compared with that obtaining at the principal S phase. In fact HOTTA, ITO, and STERN (1966) have estimated it at 0.3%. The synthesis of such a small fraction of the total DNA could not be demonstrated in this material by autoradiographic techniques. The DNA synthesized during meiotic prophase had a lower average molecular weight than that of the total DNA and a higher GC content (Table 9). Finally DNA prepared from somatic nuclei hybridized with prophase-labelled DNA to the same extent and with the same kinetic characteristics as it did with DNA labelled in somatic tissue or with DNA labelled during the pre-meiotic S phase (STERN and HOTTA 1967). Clearly this DNA is not unique to meiotic cells.

(ii) If inhibitors of DNA synthesis are added to cultured cells of *Lilium* at different stages of meiosis they interfere with the meiotic cycle only if administered during zygotene and/or pachytene (ITO, HOTTA, and STERN 1967). Zygotene arrest occurs if DNA synthesis is inhibited at the time of its initiation. Fragmentation of prophase chromosomes occurs if DNA synthesis is inhibited during mid-zygotene, while inhibition during late zygotene or early pachytene results in fragmentation during second division.

To some extent this biochemical work is paralleled by the demonstration that spot labelling with ³H-TdR appears to take place during pachytene of male meiotic prophase in the newt *Triturus viridescens* (WIMBER and PRENSKY 1963). These authors suggest that this minor synthesis is related to crossing-over. This is not, however, supported by the fact that there appears to be a fairly general distribution of grains over the chromosomes at first meiotic metaphase with no obvious relationship to the distribution of chiasmata which are distally localized in the males of this species *.

To what extent the separate replication of a small portion of the total DNA during early prophase-I is responsible, or even permissive, for the events of synapsis and/or chiasma formation is thus a matter for debate. Deoxyadenosine (ADR) inhibits DNA synthesis preferentially and meiotic cells of *Lilium* respond to ADR only during the interval of zygotene-pachytene when, of course, DNA synthesis is occurring. Moreover, cells exposed to this agent during early zygotene do not develop beyond this stage, which suggests that cells do not become fully committed to meiosis until they reach the end of prophase synthesis (STERN and HOTTA 1967). This finding is contrary to the claim of SWANSON and YOUNG (1965) that the meiotic cell once formed is irrevocably committed and can neither be induced to revert to a premeiotic state or deterred from completing its role, however aberrant the products might be.

---

* See note 2 Appendix.

In fact, STERN and HOTTA (loc. cit.) suggest that ADR interferes with the initiation of meiotic pairing, this being the most obvious of the events occurring at the time of treatment. In support of this view they quote the observations of ROTH and ITO (1967) who found that synaptinemal complexes (SC's), which are readily identifiable in control cells, cannot be detected in cells treated with ADR prior to the initiation of pairing. In microsporocytes of *Trillium* cultured from preleptotene anthers, zygotene and pachytene stages appear which are similar to those of control cultures explanted during meiotic prophase. But no chiasmata form at diplotene (desynapsis). A normal SC is present, however, at zygotene-pachytene in these desynaptic cells (ROTH and PARCHMAN 1968). The same authors also report that if protein synthesis in the lily is inhibited at late zygotene with cycloheximide, there is no visible effect on either synapsis or SC formation.

Table 9. *Base Composition of DNA Types in Lilium longiflorum.*
(Data of HOTTA, ITO and STERN 1966.)

DNA-type	Base Composition			
	C	G	T	A
1. Pre-meiotic S	20	22.8	30.2	27.0
2. Meiotic S	24.8	25.0	26.2	24.0

But again no chiasmata form and univalents are present at diplotene. They argue, therefore, that chiasma formation depends on protein synthesis, in addition to the formation of the SC. This is supported by the biochemical studies of HOTTA, PARCHMAN and STERN (1968) on protein synthesis in meiotic cells of lily following exposure to cycloheximide. At concentrations in excess of $2 \mu g$/ml, protein synthesis was virtually abolished and meiotic development completely arrested. In the range 0.2–1.0 μg/ml inhibition was incomplete and meiosis only partially suppressed, the precise cytological consequences depending on the stage and the duration of exposure. Selective inhibition of protein synthesis at the end of zygotene causes a failure of chiasma formation.

One prominent biochemical consequence of exposure to cycloheximide is an arrest of DNA synthesis. HOTTA et al. have concluded, therefore, that the limited DNA synthesis during zygotene-pachytene is itself dependent on a simultaneous synthesis of certain nuclear proteins and that the combination of these syntheses is essential for chiasma formation.

This argument is, however, not without complication for the status of the tripartite structure known as the SC remains something of an enigma. Since it is generally present at the time of synapsis and, more significantly, is located between paired homologues, it has been widely regarded as a device for securing a site by site synapsis prior to crossing over. MEYER (1960), on the other hand, has claimed that the SC is concerned with chiasma formation rather than pairing. Working with *Drosophila melanogaster* he found tripartite structures in females, where chiasma formation occurs, but not in males where, although pairing is present, meiosis is achiasmate.

l'urther, in· females homozygous for the locus C 3 G, which completely
eliminates meiotic crossing over, SC's are again absent. GASSNER (1968)
has recently reported the presence of SC's in both sexes of *Panorpa nup-
tialis* although male meiosis in this species is achiasmate *.

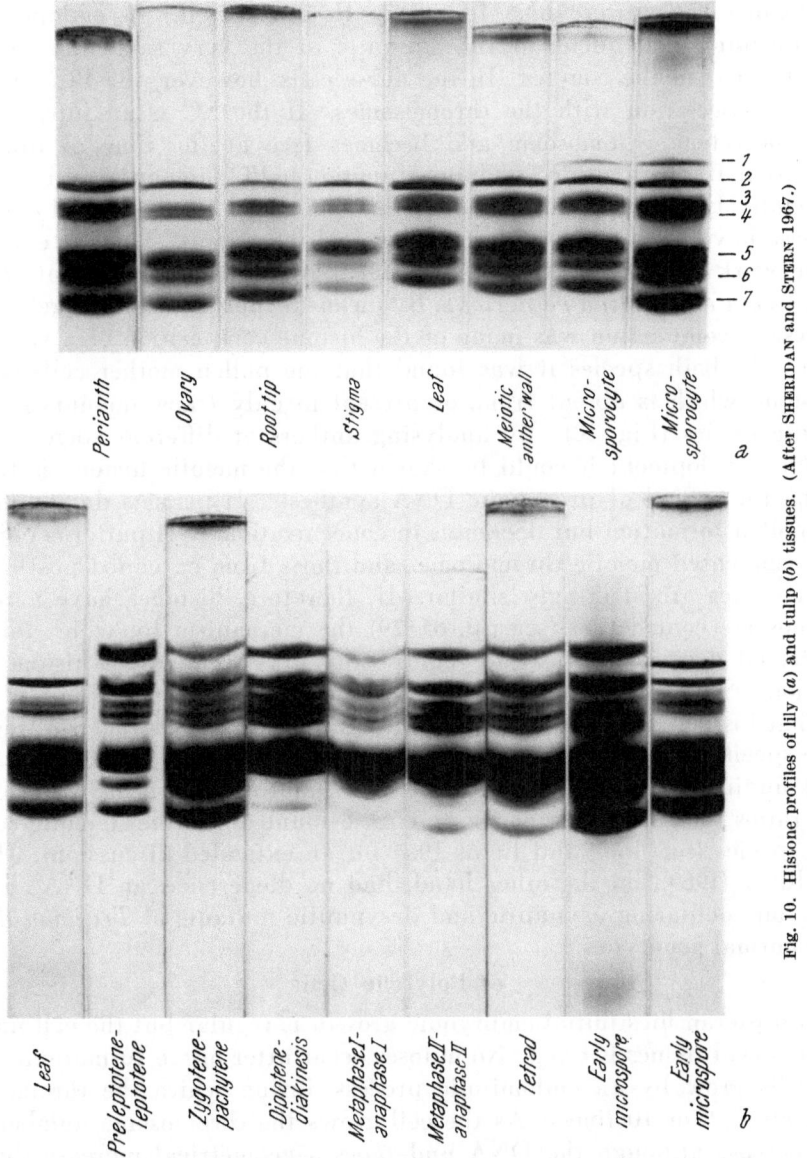

Fig. 10. Histone profiles of lily (*a*) and tulip (*b*) tissues. (After SHERIDAN and STERN 1967.)

The findings of ROTH (1966) himself complicate the issue still further.
At diplotene in primary oocytes of the mosquito *Aedes aegypti* a poly-
complex (PC), consisting of an aggregate of SC's, is regularly formed.
Initially this PC is in contact with the chromosomes but eventually it

---

* See note 3 Appendix.

comes to lie free in the nucleoplasm. A single PC may contain as many as 60 SC's but the fate of these PC's has not been followed. Comparatively simple PC's, consisting of up to seven SC's have also been reported in early spermatids of *Gryllus argentinus* and *Blaptica dubia* but their origin has never been clarified (SOTELO and TRUJILLO-CENÓZ 1960). But of greater significance, PC's were also found by ROTH (loc. cit.) in endopolyploid somatic nurse cell nuclei of the mosquito at the very same time as they were present in the oocytes. In the nurse cells, however, the PC's have no obvious association with the chromosomes. If the SC is an integral part of a pachytene chromosome and becomes free at the time of diplotene separation, it is difficult to attribute a function to PC's in non-meiotic nuclei *.

Relatively few analyses of meiotic histones have been reported and most of these have been made using cytochemical techniques. The one exception is the work of SHERIDAN and STERN (1967) in the liliaceous plants *Lilium longiflorum* and *Tulipa gesneriana*. By means of polyacrylamide gel electrophoresis a comparison was made of the histone composition of a variety of tissues. In both species it was found that the pollen mother cells contain a histone which is absent from, or present in only trace amounts in, their somatic tissues (Fig. 10). By analysing anthers at different stages of premeiotic development it could be shown that the meiotic histone is formed during the period of premeiotic DNA synthesis. It persists during meiosis and pollen formation but decreases in concentration. Gel patterns obtained from condensed meiotic chromosomes and those from extended post-meiotic chromosomes are strikingly similar. If, therefore, histones have a role in chromosome contraction (see pp. 65—79) the mechanism by which this role is effected does not appear to require any gross changes in histone composition.

That histones may be important in determining the character of meiosis, more specifically the occurrence of synapsis, is suggested by ANSLEY's (1954, 1968) findings that asynaptic cells in *Loxa* and *Scutigera* have DNA: histone ratios that depart from the 1 : 1 ratio found in normal meiotic cells of both species (see JOHN and LEWIS 1965 for an extended discussion). HAYTER and RILEY (1969), on the other hand, find no dienerences in DNA : histone ratios on comparing asynaptic and desynaptic mutants of *Triticum durum* with normal genotypes.

c) Polytene Cells

In dipteran flies initial embryonic growth is regular but the cell number in the larval tissues is fixed. No mitoses occur after organ formation; rather the cells grow by an endomitotic process during which the chromosomes replicate 8, 9 or 10 times. As the cell grows the chromosome number does not increase although the DNA undergoes a geometrical increase (but see p. 38). The polytene system is thus the result of successive endomitotic divisions in which the products of replication remain together. This, coupled with the retention of somatic pairing, results in homologous chromatids being arranged in the form of a single chromosome. The number of strands

---

* See note 4 Appendix.

Table 10. *Distribution of Polytene Nuclei in Larval and Prepupal Salivary Glands of Drosophila melanogaster. The class number represents the Level of Replication.*
(Data of RODMAN 1967.)

Sex	Stage	Age (hrs.)	No. Nuclei	3	3–4	4	4–5	5	5–6	6	6–7	7	7–8	8
Female	Larvae	72	34	2.5	3	44	53	7						
Female	Larvae	78–80	41		22	14.5	53.3	2						
Female	Larvae	75–82	48		4	2	73	3						
Female	Larvae	114–117	30				7		19	13.5	6.5	5	36	17
Female	Larvae	117–119	42						76	5	15	8	46	
Female	Larvae	120–122	35						32	11	14			
Female	Prepupal	4	49						3	4	14	35	41	4
Female	Prepupal	6	52						2		5	50	42	2
Female	Prepupal	4–6	50							4	12	30	28	26
Female	Prepupal	6–8	34							9	3	27	29	26
Female	Prepupal	7½	44							4	13	25	32	27
Male	Larvae	119–121	58						5	7	24	5	59	
Male	Prepupal	4–6	50						14	8	40	4	34	
Male	Prepupal	7½	56						6		5	38	55	2
Male	Prepupal	9	52								12	10	44	29

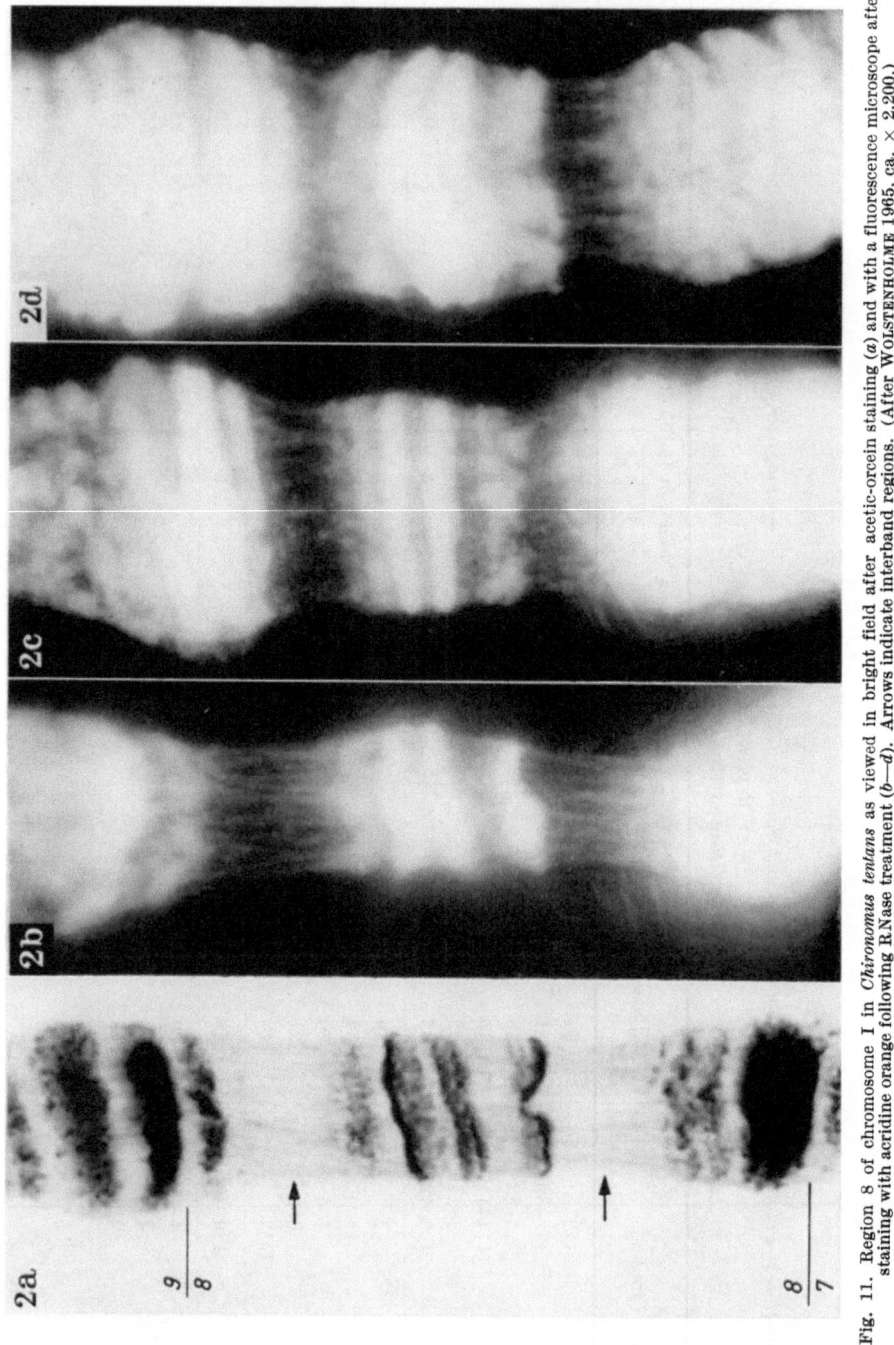

Fig. 11. Region 8 of chromosome I in *Chironomus tentans* as viewed in bright field after acetic-orcein staining (*a*) and with a fluorescence microscope after staining with acridine orange following RNase treatment (*b—d*). Arrows indicate interband regions. (After WOLSTENHOLME 1965, ca. × 2,200.)

in a polytene chromosome varies from tissue to tissue but the highest number is usually attained in the salivary glands of the prepupal stage.

For example in *D. melanogaster* most of the salivary gland nuclei have already completed their 4th replication by the 2nd instar (72 hour larva) and some have almost completed the fifth replication at this stage. The initiation of the final (8th) replication cycle takes place within a short period just before the last 5 hours of the 48 hr. third instar and continues throughout the entire prepupal period (RODMAN 1967, Table 10). Previous references to the classes of polytene nuclei found in the larval salivary glands of *D. melanogaster* are those of SWIFT (1962), who estimated the highest class as 1024× corresponding to nine replications of the diploid

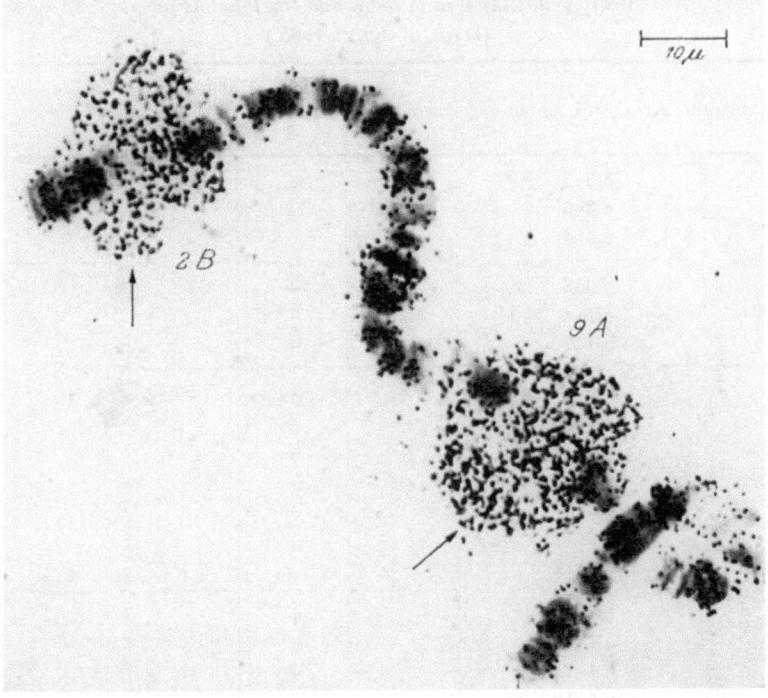

Fig. 12. Chromosome II of *Sciara coprophila* following 15 min incubation with ³H-TdR to show puffs at maximum extension. (After GABRUSEWYCZ-GARCIA 1964, ca. ×1,080.)

nucleus, and of RUDKIN (1963) who notes the highest polytenic replication is the 8th. As RODMAN (1967) points out, neither of these authors indicate how these values were obtained.

Two features which accompany replication are increasing length and diameter, on the one hand, and the development of a system of transverse banding based on the occurrence of chromatic zones separated by achromatic interbands, on the other. There has long been a controversy as to whether the interband regions contain DNA. WOLSTENHOLME (1965, 1966) has provided convincing evidence from fluorescence microscopy that they do (Fig. 11). In addition DNase disrupts the interband areas of polytene chromosomes (LEZZI 1965) while RNase and proteases do not. Moreover, histone, as well as DNA, is found in the interbands (GOROVSKY and WOODARD 1967).

Their extended length, lateral multiplicity and transverse banding make the polytene chromosome a particularly favourable system for studying the metabolic patterns of specific chromosome sites. They offer levels of resolution with, and sensitivity to, microchemical methods which cannot be approached with conventional somatic chromosomes. Further, local variations occur in the clarity of banding both in different tissues and at different times in the same tissue. These variations are related to the functional activity of the band which undergoes a process termed puffing. Bands first

Table 11. *Comparative DNA Ratios of Twenty Equivalent Loci in F_1 Hybrids between Chironomus thummi thummi and Ch. thummi piger.*
(Data of KEYL 1965.)

Chromosome Arm and Locus		DNA Ratio (*thummi/piger*)				
		1	2	4	8	16
IR	c 3–3		+			
	c 3–6	+	+			
	c 4–1			+		
IL	d 3–8	+	+			
	e 1–4		+			
	e 1–7		+			
IIR	b 5–28	+	+	+		
	c 1–7	+	+			
	c 1–8	+	+			
	c 1–11		+	+		
	c 2–3	+	+			
	c 2–11		+			
	c 2–12			+		
IIL	c 4–3		+			
	c 4–6			+		
	c 4–7		+	+		
	c 4–12		+			
III	b 1–13				+	+
	b 3–11		+	+	+	+
	b 3–17		+			

become less dense and then give rise to swollen regions (Fig. 12) which may also destroy the structural integrity of neighbouring bands. The extent of puffing is variable; it takes its most pronounced form in the Balbiani rings of Chironomids which are very large, specialized puffs.

The topography of DNA synthesis in polytene chromosomes can be studied autoradiographically by incubating excised salivary glands in a medium containing ^{3}H-TdR. In *Chironomus* autoradiographs of nuclei fixed a short time after ^{3}H-TdR treatment show two main types of labelling. In the one case almost all parts of the chromosome are labelled while in the other the label is confined to specific, well-defined bands (KEYL and PELLING 1963). These two patterns, in fact, represent the extremes of a series which represents one total replication cycle. By double labelling it was found

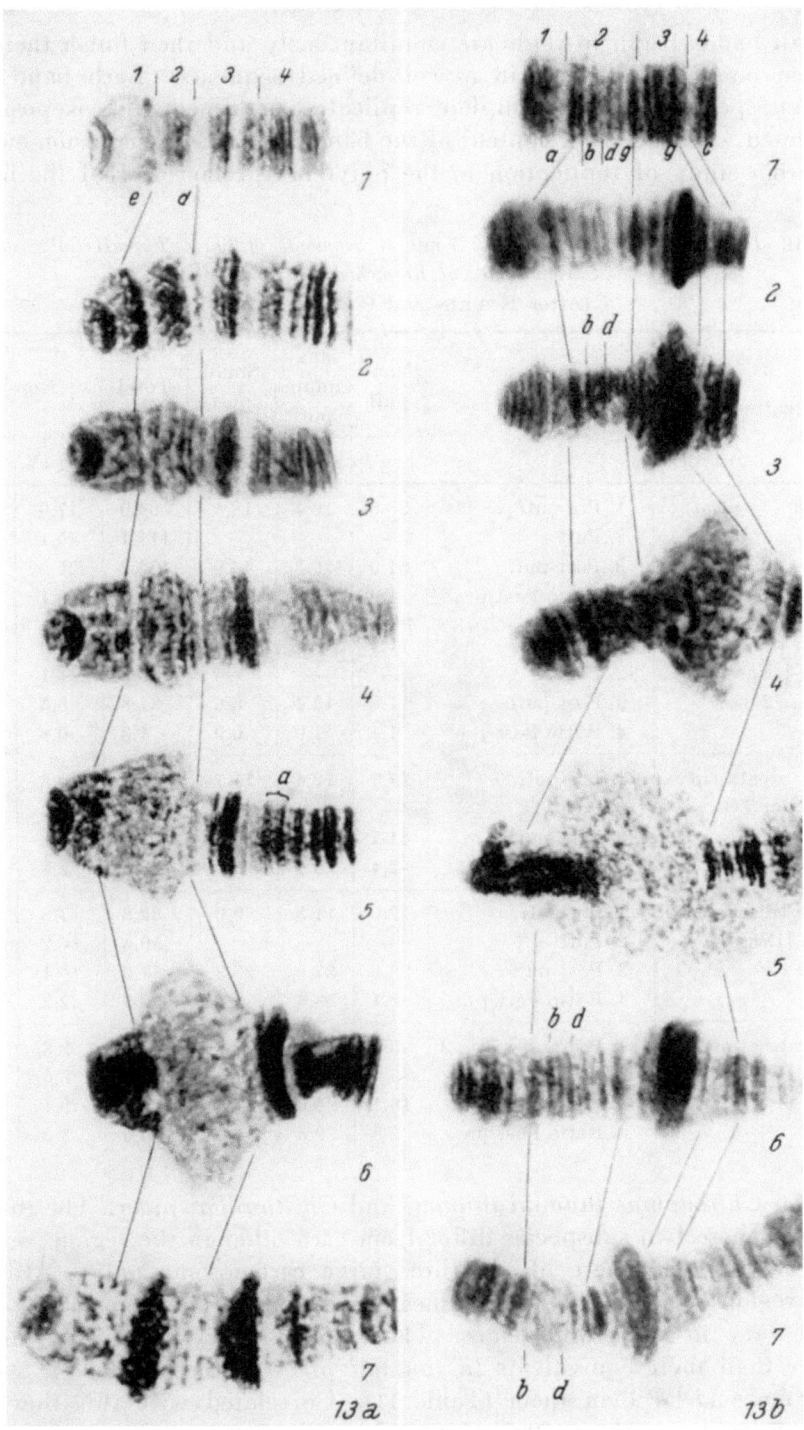

Fig. 13. Behaviour of DNA puffs in salivary gland chromosomes of *Rhynchosciara angelae*. Series (*a*) illustrates the distal end of chromosome B and shows the behaviour of the B 2 band which accumulates DNA during bulb formation and then retains it after regression of the bulb. Also shown is band 3 in which the accumulation of extra DNA precedes bulb formation and is maintained after it. Series (*b*) shows the distal end of chromosome C. Here band 3, like its counterpart B 3, accumulates DNA before bulb formation and retains it throughout and after the process. (After BREUER and PAVAN 1955.)

that all bands begin to replicate simultaneously and then finish their replication one after another in a well-defined sequence. Each band takes
its own specific time to complete replication, a time which is positively
correlated with the DNA content of the band. The same conclusion emerges
also from study of replication in the polytene chromosomes of the hybrid

Table 12. *Integrated Absorbancies at 257 mμ of Segments of Three Type-B Salivary Gland Chromosomes of Rhynchosciara angelae.*
(After Rudkin and Corlette 1957.)

Treatment	Stage	Non-puff	Max-imum puff	Small puff	Total	Non-puff	
		Seg I	Seg II	Seg III	I–III	Seg IV	Seg VI
1. Before treatment	1. Pre-puff	26.5	19.4	18.1	64.0	17.0	52.5
	2. Puff	—	—	—	111.1	25.1	77.1
	3. Post-puff	51.0	76.3	44.9	172.2	35.0	106.3
	4. Ratio Post/pre	1.9	3.9	2.5	2.7	2.1	2.0
2. After treatment with RNase, DNase & hot TCA	1. Pre-puff	11.1	6.9	7.0	25.0	8.0	23.7
	2. Puff	—	—	—	29.3	4.1	18.3
	3. Post-puff	11.8	13.4	6.6	31.8	6.6	24.8
	4. Ratio Post/pre	1.1	1.9	0.9	1.3	0.8	1.0
3. After treatment with hot TCA	1. Pre-puff	17.7	12.4	11.7	41.8	10.3	24.6
	2. Puff	—	—	—	85.4	22.8	44.1
	3. Post-puff	41.7	62.8	35.1	139.6	25.2	63.9
	4. Ratio Post/pre	2.4	5.1	3.0	3.4	2.4	2.6
4. After treatment with DNase	1. Pre-puff	12.0	11.3	9.0	32.3	7.5	21.8
	2. Puff	—	—	—	50.6	14.2	38.6
	3. Post-puff	25.6	37.9	23.3	87.0	16.1	41.2
	4. Ratio Post/pre	2.1	3.4	2.6	2.7	2.2	1.9
5. After treatment with RNase	1. Pre-puff	5.7	1.1	2.7	9.5	2.8	2.8
	2. Puff	—	—	—	34.8	8.6	5.5
	3. Post-puff	16.1	24.9	11.8	52.8	9.1	32.7
	4. Ratio Post/pre	2.8	22.6	4.3	5.6	3.3	11.7

between *Chironomus thummi thummi* and *Ch. thummi piger*. The chromosomes of these two subspecies differ from each other in the regions on both
sides of the centromere of the three large chromosome pairs. Although
these regions correspond in the linear arrangement of their bands, these
bands vary in their dimensions. Those in *thummi thummi* are always
thicker than their equivalents in *thummi piger* and contain 2, 4, 8 or 16
times more DNA than them (Table 11). Correlated with this they take
longer to replicate (Keyl 1965, 1966).

In section 2 of chromosome B of *Rhynchosciara angelae* one locus swells
enormously, accumulates DNA and then returns to a banded stage which
maintains the extra DNA until the histolysis of the tissue (Breuer and

PAVAN 1955). In section 3 of chromosome B and section 3 of chromosome C disproportionate synthesis of DNA precedes the formation of the bulb and is maintained during and after it (Fig. 13). It is still not clear whether the extra DNA is intercalated into the chromatids or whether it represents extra material replicated in addition to the existing polytene strands. Nor is there evidence as to its fate. That the cytological picture presented by BREUER and PAVAN does depend on a disproportionate increase in DNA

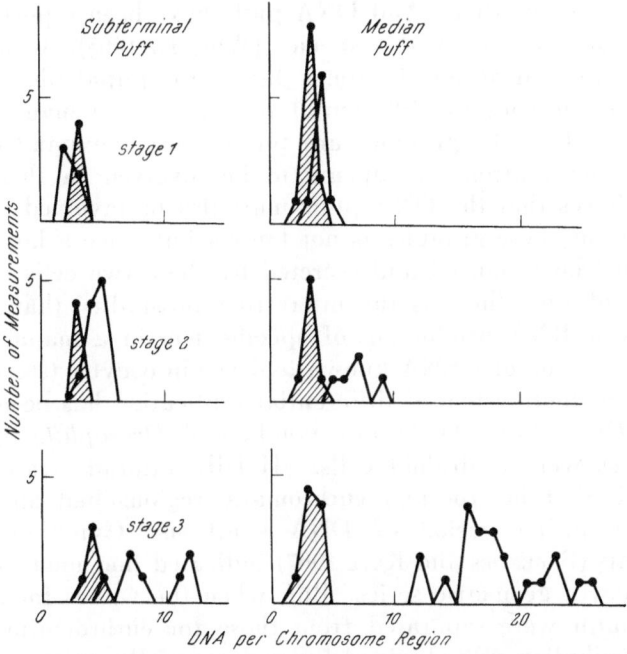

Fig. 14. DNA content of individual puffs (unshaded regions) and adjacent unpuffed sections (dotted regions) in chromosome II of *Sciara coprophila*. All measurements are from chromosomes of the 2048 C class. Stage 1 = late fourth instar larva, Stage 2 = early prepupa, and Stage 3 = late prepupa. (After SWIFT 1962.)

has been demonstrated by the measurement of the quantity of UV-absorbing material present in the regions concerned (RUDKIN and CORLETTE 1957). For this purpose they divided the distal portion of chromosome B into six adjacent segments in two of which (II and III) puffing occurred. Their measurements show that the DNA in the puff region at least doubles with respect to non-puffing regions (Table 12).

Sciara coprophila also shows localized increases in the DNA (SWIFT 1962). Here they occur in at least four specific regions of the chromosome complement beginning just before pupation. At this time the salivary glands become bloated with a muco-protein cement secretion used for attaching the developing pupal case. By measuring the DNA content of individual puffs and adjacent regions SWIFT showed that the DNA content of certain puffs is augmented (Fig. 14) *.

Comparable DNA-synthesizing puffs have not been found in *Chiro-*

---

* See note 5 Appendix.

nomus or in *Drosophila.* Schultz (1965) has therefore suggested that where a nucleolar organiser functions to maintain a single compact nucleolus the extra, disproportionate DNA synthesis may not be required. Thus in *Rhynchosciara* and *Sciara* where the DNA puffs are established there is no organized nucleolus of the kind found in *Chironomus* and *Drosophila.*

DNA puffs have been found also in the polytene chromosomes of the pupal foot pad cells of dipterans. In *Sarcophaga bullata,* for example, there are two giant cells per pad and these serve to secrete the cuticle of the adult fly during pupation. And DNA puffs have been reported on at least two of the chromosomes in this species (Whitten 1965). In addition DNA granules are present at specific times during the pupal phase. These arise from the chromosomes by differential replication. In many instances the regions from which the granules are formed are unexpanded. Indeed, in some cases, the centromere appears to be involved in their production. Whitten believes that the DNA puffs may also be involved in the process. The function of these granules is not known but since a large quantity of material must be produced and secreted by these two cells in a relatively short space of time there is the interesting possibility that they serve to amplify the m-RNA production of specific loci in a manner reminiscent of the amplification of r-RNA known to occur in oocytes (see p. 56).

A quite distinct system of differential replication has been described in species of *Drosophila.* The brain ganglion of *Drosophila hydei* contains "polytene" as well as diploid cells. ^{3}H-TdR autoradiography of diploid cells showed that hetero- and euchromatic regions had an asynchronous though overlapping period of DNA synthesis. Cytophotometric DNA measurements (Berendes and Keyl 1967) indicated that nuclear DNA values did not follow a geometric series. But, when the values for chromocentral heterochromatin were separated from those for euchromatin, a good geometrical distribution was obtained for each type. Berendes and Keyl therefore suggest that the process of polytenisation involves independent replication of eu- and heterochromatin and that the number of replication cycles completed by each can be different even within the same cell. Rudkin (1963, 1965), working with *D. melanogaster,* reports that most of the heterochromatin does not replicate at all during the formation of polytene chromosomes in the salivary gland.

d) Endoploid Cells

Ciliate Protozoa have two morphologically and functionally different kinds of nuclei. The micronucleus is small, divides by mitosis or meiosis and engages in little if any RNA synthesis. The macronucleus is large, divides amitotically and has distinct nucleoli and an active RNA synthesis. The macronucleus originates from a diploid anlage, one of the mitotic products of the fusion nucleus produced by conjugation. The development of the macronucleus is associated with a great increase in nuclear DNA content. This transformation is not, however, simply one of ploidy, it is also associated with marked changes in the structural organization of the chromatin. These changes are best understood in *Stylonychia mytilus*

(AMMERMANN 1963). Here the macronuclear anlage initially form polytene chromosomes. This state is, however, only temporary. Transverse breakage between the dense, DNA rich bands results in the formation of a very large number of chromatin granules each consisting of a band region.

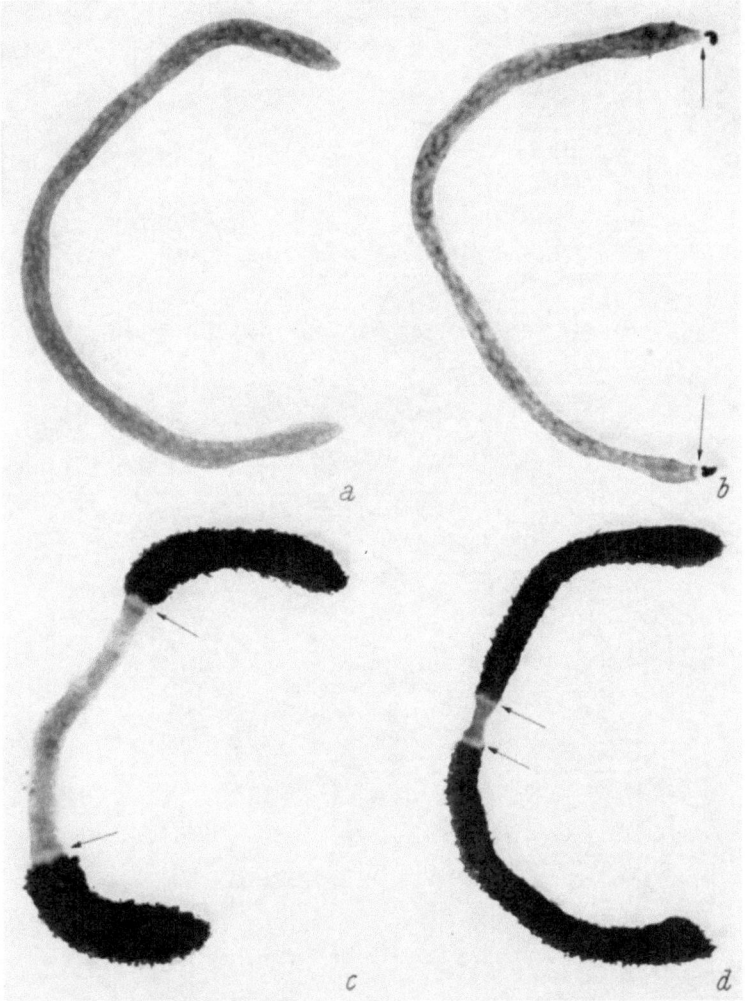

Fig. 15. Isolated macronuclei of *Euplotes eurystomus* labelled with ³H-TdR to illustrate the behaviour of the replication bands (arrows). *a*—From a cell in G_1 (ca. 25% through the cell cycle). *b*—Onset of S (ca. 30% through the cell cycle). *c*—Mid S (ca. 60% through the cell cycle). *d*—Late S (ca. 90% through the cell cycle). (After PRESCOTT 1963, ca. ×1,000.)

In *Euplotes* the macronucleus has the form of a long thread. During interphase a profound reorganization of the macronucleus occurs. Two clear transverse zones, known as reorganization bands, appear at the extreme ends of the macronucleus. These bands move progressively toward the centre of the macronucleus (Fig. 15). Here they meet, fuse and then disappear shortly before the macronucleus divides amitotically by being

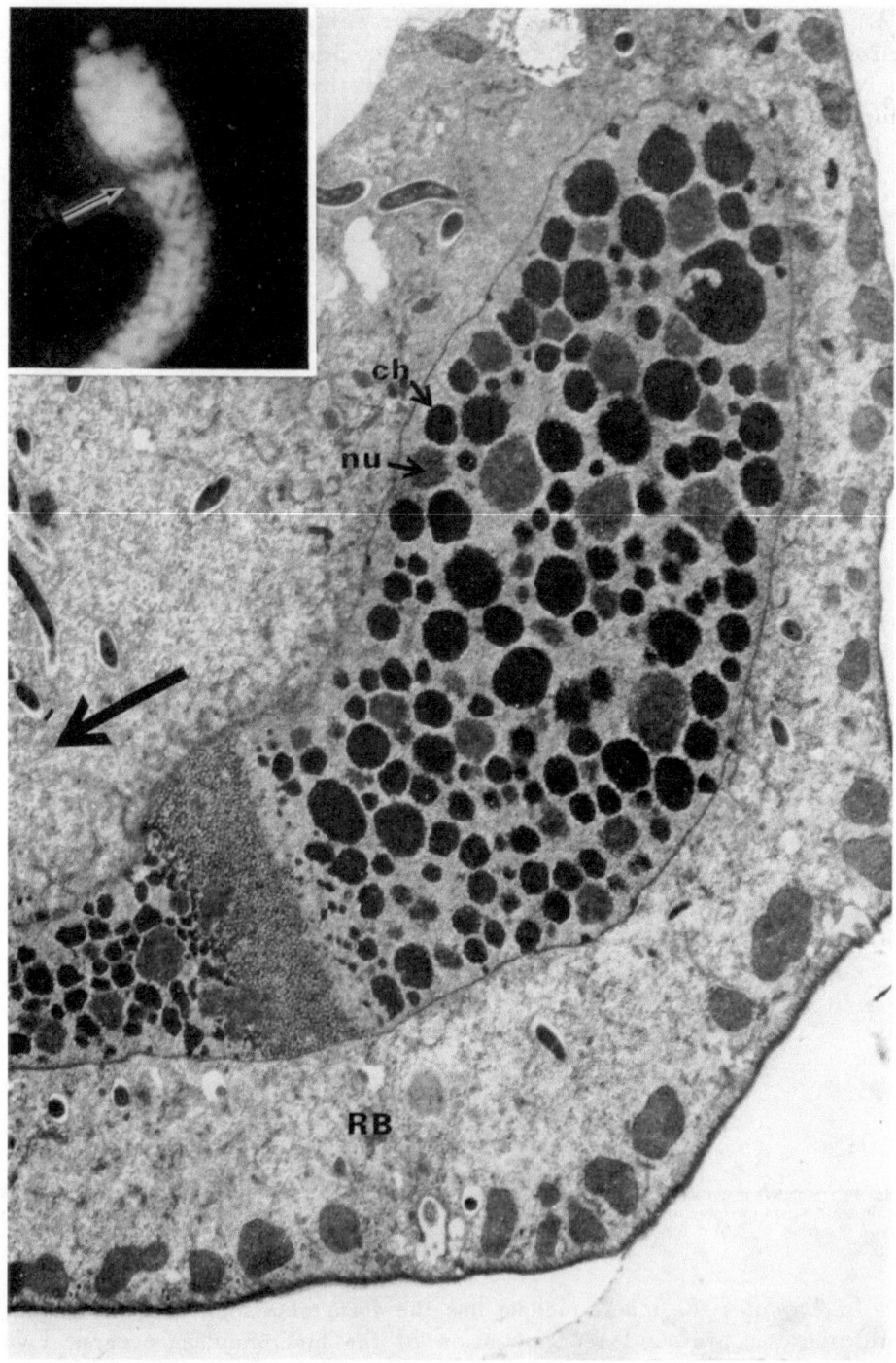

Fig. 16. Organization of the macronucleus of *Euplotes* as seen with the electron microscope (*a*—ca. ×7,000). The reorganization band (RB) is moving in the direction indicated by the arrow. ch = chromatin granules, nu = nucleoli. Note that the pre-reorganization granules (left of RB) are considerably smaller than their post-reorganization counterparts (right of RB). Also shown (*b*) is a fluorescence photomicrograph of an acridine orange stained macronucleus. (After RINGERTZ, ERICSSON, and NILSSON 1967.)

pinched off at the middle. GALL (1959) first demonstrated that the reorganization bands are zones of intense DNA synthesis and no disjunct labelling occurs between them. Histone synthesis too occurs strictly at the site of DNA synthesis in the replication band. This system thus offers a novel approach to the problem of the chromatin changes that occur during DNA replication.

The macronucleus of *Euplotes* contains approximately one hundred times as much DNA as the micronucleus. Ordinarily the macronuclear chromatin is organized into distinct rounded granules which vary in size. The nuclei also contain numerous nucleoli. In the advancing edge of the

Table 13. *The Consequences of Nuclear Transference at Different Phases of the Cell Cycle in Stentor coeruleus.*
(After DE TERRA 1967.)

	Transfer Type		Distribution of ^{3}H-TdR Label
1.	(a) G_1-nucleus into S-cell		Label over both host and transfer nuclei
	(b) G_1-nucleus into G_1-cell		No label over host or transfer nucleus
2.	(a) S-nucleus into G_1-cell		No label over host or transfer nucleus
	(b) S-nucleus into S-cell		Label over both host and transfer nuclei
3.	(a) Late D-nucleus into S-cell		Label over both host and transfer nuclei
	(b) Late D-nucleus into G_1 cell		No label over host or transfer nucleus.

reorganization band the chromatin granules are replaced by fibrils (Fig. 16). At the distal face of the reorganization band the fibrils re-aggregate into chromatin granules, the average volume of which is six to eight times that of the pre-organization granules (RINGERTZ, ERICSSON, and NILSSON 1967). These post-organization granules must of course decrease in size after amitosis and prior to the next S phase. At amitosis of the macronucleus of *Tetrahymena pyriformis* CLEFFMANN (1968) has shown that the distribution of DNA between sisters is not equal. Two compensatory processes then serve to regulate the mean DNA content of the macronucleus: (i) The elimination of DNA-containing bodies at macronuclear division, the precise DNA content of these bodies being positively correlated with the DNA content of the macronuclei from which they arise, and (ii) An additional replication which takes place in the generation immediately following the production of a macronucleus with a lower limit of DNA.

The large heterotrich ciliate *Stentor coeruleus* contains many micronuclei and one large polyploid macronucleus which throughout interphase exists as a chain of nodes running almost the entire length of the organism. At 20° C amitotic division takes approximately eight hours and the macronucleus undergoes a series of eight striking morphological changes each occupying a one hour period. At stage 5 of this process the nuclear nodes begin to coalesce until the nucleus forms a compact mass in the centre of the cell (stage 6). This subsequently elongates into a thin rod (stage 7) which forms nodes again as it is passively pinched into two by the cleavage furrow (stage 8).

By culturing dividing organisms (stages 3–8) in a ^{3}H-TdR containing medium for 30 min and then making autoradiographs of the organisms, DE TERRA (1967) found that label was always present above the nuclear nodes at stages 3 and 4, seldom present above coalescing nuclei (stage 5) and never present above the compacted, elongating and nodulating nuclei (stages 6–8). DNA synthesis thus occurs during the first four hours of the 8 hour division period (D) and stops at or near the time nuclear coalescence begins.

In *Stentor*, as in *Euplotes*, the macronuclear cycle is thus:

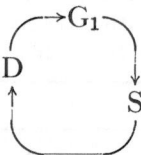

If the nuclei of cells in D and G_1 are transferred to cells in S, DNA synthesis is initiated in them (Table 13). DE TERRA therefore concluded that macronuclear DNA synthesis in *Stentor* is determined by a cytoplasmic factor. This factor is an initiator present in S cells rather than an inhibitor present in those of D and G_1 since when S and G_1 cells of equal size are grafted together S is initiated in the G_1 cell and not inhibited in the S cell. Finally, the initiator is required throughout synthesis since nuclei from cells in S stop synthesis when transferred to cells in G_1.

e) Diploid Somatic Cells

The incorporation of labelled precursors into DNA has generally been regarded as representing either pre-division synthesis or else an endomitotic replication leading to states of endoploidy or polyteny. PELC (1964) has, however, argued for a third category whereby incorporation leads neither to nuclear division nor to endomitotic replication. Rather it is involved in DNA renewal or replacement, a process which PELC has termed replacement synthesis. The evidence for such synthesis is based on the fact that while, in some organs (intestines and oesophagus), all ^{3}H-TdR labelled cells divide, in others more cells become labelled than subsequently divide. For example, in seminal vesicles, heart muscle and smooth muscle of adult mice the incorporation of labelled thymidine exceeds the requirement for cell division by a factor of 8–12.

PELC concludes, therefore, that cell rejuvenation can occur in differentiated cells of mammals by a process other than cell division. This involves the incorporation of DNA precursors without any subsequent division. What is not clear is whether this incorporation represents a complete re-synthesis of DNA with a temporary increase to the 4 C level or whether the breakdown of DNA occurs with or without preceding synthesis. One possibility suggested by PELC is that, following separation of the DNA strands in preparation for synthesis, only one strand replicates, the other being discarded. The net result would then be the maintenance of a 2 C-state. Alternatively part of the DNA of a nucleus might be renewed by a similar mechanism. Indeed, it seems conceivable that DNA molecules become unusable and require replacement.

This claim by PELC is contrary to the assumption that the DNA of a nucleus is stable and that its synthesis takes place only during chromosome duplication. One other exception to this assumption has already been referred to, namely the occurrence of additional DNA in puffs of *Rhynchosciara* and *Sciara* (see p. 37). A further example involving the augmentation of nucleolar DNA in the lampbrush chromosomes of female newts will be dealt with later (see pp. 56–57).

Table 14. *Distribution of Metabolically Labile DNA of Wheat.*
(Data of SAMPSON et. al. 1963.)

Tissue	Proportion of Low Molecular Wt. DNA
1. Male germinal	0%
2. Growing regions of root and leaf	up to 20%
3. Dormant embryo in wheat seeds	ca. 10%

An alternative method of accommodating these variations would be to assume that there are two types of DNA—a stable kind with a strict genetic function and a labile one with only metabolic functions. Metabolic DNA would then be synthesized when cells require an amplification of specific metabolic information. In its simplest form this might be expected to lead to a copying of existing DNA to give a metabolic equivalent with similar properties. SAMPSON et al. (1963), however, have drawn attention to the fact that DNA prepared from a variety of growing plant tissues can be resolved on a methylated albumin column into two distinct fractions. Both show melting profiles typical for a double stranded helix. But viscometric analysis indicates that their respective molecular weights are $2-3 \times 10^5$ and $4-6 \times 10^6$. The proportions of the two forms varies with the type of tissue (Table 14) and even with the physiological state of a particular tissue.

High molecular weight DNA has the same composition irrespective of the tissue concerned and no evidence of turnover in this DNA was obtained. This form of DNA thus behaves in the manner expected of genetic material. This does not appear to be true of the low molecular weight form which has a relatively rapid rate of turnover and a composition quite different from that of its high molecular weight counterpart.

What relationship this metabolic DNA bears to so-called satellite DNA (SKINNER 1967) is not clear. SKINNER's work shows that in the mouse and the crab *Cancer borealis* a significant fraction of the DNA forms distinct and separate bands in a CsCl density gradient centrifugation. Similar satellite DNA's, comprising from 10–30% of the total DNA, have also been found in seven other species of *Cancer*.

2. Chromosome Metabolism

a) Mitotic Cells

Cellular RNA is synthesized mostly, perhaps entirely, in the nucleus, both on the chromosomes and in the nucleolus. It is now clear that the chromosomes in interphase, and in early and mid-prophase, function in RNA synthesis. It is generally claimed, however, that they cannot do so when they are in their condensed meta- and anaphase form (Fig. 17). For example

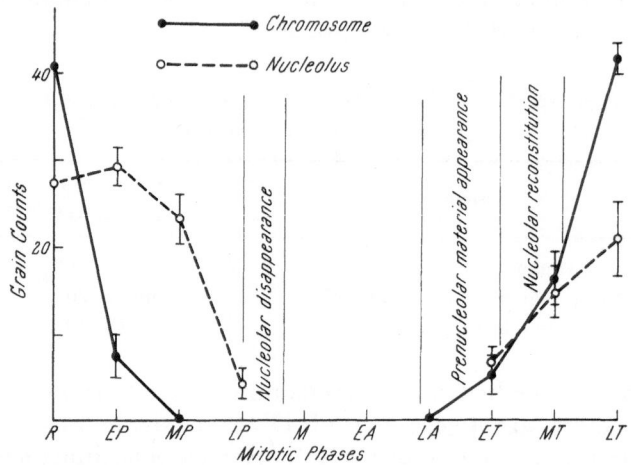

Fig. 17. Average grain counts of chromosomes and nucleolus of *Luzula purpurea* at different stages of the mitotic cycle. Data taken from autoradiographs of root tip cells after treatment with ^{3}H-UdR (a. 5 c/ml) for 10 min. The ana-telophase (A and T) data are the sum of counts over chromosome groups. (After Kusanagi 1964.)

Davidson (1964) reports that late prophase cells of *Vicia faba* roots no longer synthesize RNA and that which was synthesized in earlier stages is lost to the cytoplasm at a rapid rate. This loss appears to be related in part to the breakdown of the nuclear membrane. But it may also depend upon the fact that active chromosome contraction in late prophase precludes the uptake of RNA by the chromosomes.

Now ^{3}H-adenosine is a more effective label for chromosomal RNA than ^{3}H-cytidine because the chromosomal RNA has a higher adenine content than nucleolar RNA. Using ^{3}H-adenosine La Cour (1963) showed that RNA is present on the metaphase chromosomes in root meristems of *Trillium cernuum*. With the same technique it was shown that the nucleoli incorporated the labelled precursor during prophase and that aggregations of labelled material appeared on the chromosomes in the vicinity of the nucleolus when it disappeared in the later stages of prophase. These observations imply that nucleolar material is transferred to the chromosomes during prophase. La Cour therefore suggests that the metaphase chromosome may be a vehicle for carrying dispersed nucleolar material from prophase to telophase. Using short-term (10 min) ^{3}H-CdR treatment Kusanagi (1964) has also claimed that nucleolar RNA moves onto the metaphase chromosomes of *Luzula purpurea* during late prophase. DAVIDSON (1964), while agreeing that some RNA is certainly present in the mitotic chromosomes at metaphase, claims that only a small fraction of the RNA synthesized in the nucleus during prophase or late interphase is incorporated

Table 15. *RNA Content of Isolated Metaphase Chromosomes.*
(Data of HUBERMAN and ATTARDI 1966.)

Item	RNA/DNA (mg)
1. Metaphase chromosomes	0.66
2. Whole nuclei	0.38
3. Interphase chromatin	0.15

into the chromosomes. On the other hand HUBERMAN and ATTARDI (1966) have presented data which point to the presence of a large amount of RNA in the metaphase chromosome compared with that in interphase chromatin or whole nuclei (Table 15).

Using whole-cell grain counts on hamster epithelium incubated with ¹⁴C-uridine, KONRAD (1963), contrary to the reports of other investigators, found that RNA synthesis during mitosis was significantly above zero (Fig. 18). LIN, KARKAS, and CHARGAFF (1966) likewise find that metaphase chromosomes from HeLa cells will support the synthesis *in vitro* of polyribonucleotides by the RNA polymerase of *Escherichia coli*. The rate of synthesis using chromosomes as a template turns out to be lower than that obtained with free HeLa DNA. With both types of template, however, the RNA formed has a high

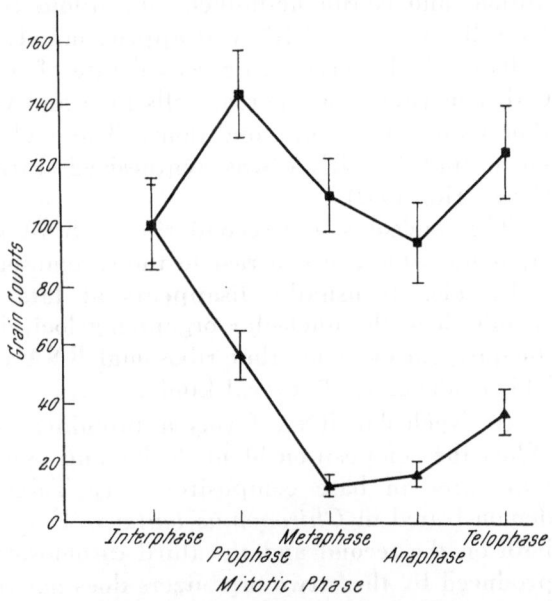

Fig. 18. Normalized average grain counts corrected for self-absorption in hamster cells following uptake of radioactive uridine (— ▲ —) and phenylalanine (— ■ —). Ninety-five percent confidence limits are included for each point. (After KONRAD 1963.)

AU content and does not resemble that of the normal RNA component of the chromosomes which has a high GC content.

CRIPPA (1966) carried out an experiment in which the labelling of newly synthesized RNA with ³H-UdR was coupled with histophotometric determination of the DNA component of the nucleus. This made it possible to correlate the rate of RNA synthesis (expressed as grain count per cell) with the change in DNA content. He reports that in a quasi-diploid Chinese hamster cell line the rate of chromosomal and nucleolar RNA synthesis is lowest in G_1, increases continuously through S and thereafter remains constant through G_2. Synthesis drops at the end of prophase and ceases at metaphase. 15% of the 500 metaphase cells counted proved to be tetraploid. These followed the same pattern though the rate of their RNA synthesis was about twice that of the diploids.

When the duration of DNA synthesis and the mitotic cycle time in a tetraploid root of *Pisum* were compared with those of cells in a diploid root, both proved to be a function of cellular DNA content (VAN'T HOF 1965). If, however, a small portion of cells in a diploid root are induced by colchicine to become 4 x the cycle time of 2 x and 4 x cells turns out to be approximately equal and the S periods are almost the same (VAN'T HOF 1967) indicating that the 4 x cells are synthesizing DNA at twice the rate of the neighbouring diploids. It is worth noting that the 4 x cells studied in the colchicine treated roots were those of the X_1 generation. Whether the relations found continue to hold for subsequent 4 x cell generations remains to be seen. Utilization of ^{3}H-CdR to measure RNA synthesis of diploid and colchicine-induced tetraploid cells likewise indicated that the 4 x cells synthesized RNA at approximately twice the rate of diploid cells. Cells in G_1 displayed a decreased rate of incorporation with increased ^{3}H-CdR concentration whereas cells in S and G_2 showed increased incorporation under the same conditions. These observations were interpreted to mean that less RNA was synthesized during G_1 than during S and G_2 (VAN'T HOF 1967).

The nucleolus, the second site of RNA synthesis in the nucleus, is an organelle which, in contrast to the chromosome, lacks continuity during the cell cycle. It usually disappears at late prophase and is reformed at specific loci, the nucleolar organizing loci, during telophase. There is now unequivocal evidence that ribosomal RNA is contained within the nucleus This evidence is of several kinds:

(i) Nucleolar RNA forms a prominent fraction of the nuclear RNA. When the composition of nucleolar and cytoplasmic RNA is compared the two agree in base composition (see Table 19). This is especially well demonstrated in *Chironomus tentans*. Here nucleolar organizers are found both on the second and the third chromosomes. Base analysis of the RNA produced by these two organizers does not reveal any significant difference between them (EDSTRÖM 1964). Moreover, RNA particles of identical size, namely 28 S and 18 S occur both in the nucleolus and the cytoplasm (see also p. 12).

(ii) The development of methods for isolating nucleoli has facilitated the localization of RNA polymerase and DNA-primed RNA in isolated nucleolar systems. For example, high-resolution autoradiography shows that the Mg^{2+}-activated RNA polymerase reaction, responsible for the synthesis of ribosomal RNA, occurs primarily in the nucleolus (MAUL and HAMILTON 1967). By contrast the $Mn^{2+}+(NH_4)_2SO_4$-activated RNA polymerase reaction that synthesizes a more DNA-like RNA occurs primarily on the chromosomes in the extranucleolar region.

(iii) The hybridisation of RNA with denatured DNA provides a tool for identifying DNA complementary to any given RNA. With this technique it has been shown that approximately 0.3% of the genome is complementary to ribosomal RNA (CHIPCHASE and BIRNSTIEL 1963, RITOSSA and SPIEGELMANN 1965). In some cases the DNA complementary to ribosomal RNA appears to be clustered at the nucleolar-organizing region. This implies that the organizer represents several thousands of gene loci clustered together which are probably transcribed as a unit. PERRY (1965), on the other hand, has argued that there are between 400 and 2,000 cistrons for r-RNA production which are dispersed throughout the genome. On this basis he assumes that the RNA is synthesized principally by extra-nucleolar chromatin and is then transferred to the nucleolus.

(iv) A recessive mutation is known in *Xenopus laevis* which prevents the formation of a normal nucleolus when homozygous (ELSDALE, FISCHBERG, and SMITH 1958). Both DNA and 4S-RNA are synthesized by the anucleolar mutant but the mutation prevents the synthesis of 28S and 18S ribosomal RNA (BROWN and GURDON 1964) so that no ribosomes form. Ultrastructure studies show that the 150—300 Å granules normally found in the nucleolus are absent in the mutant (JONES 1965). This mutation thus leads to the loss of most of the ribosomal cistrons though it is normal in respect of non-ribosomal RNA fractions.

(v) Selective irradiation of nucleoli with a UV-microbeam strongly inhibits the appearance of newly formed RNA in the cytoplasm (PERRY and ERRERA 1960).

(vi) In general only those micronuclei which maintain nucleolar bodies are able to synthesize DNA and RNA (see p. 25). That is, the ability to synthesize nucleic acid is correlated with prolonged maintenance of nucleolar material. This dependence of micro-nuclear RNA synthesis is understandable if a major fraction of nuclear RNA synthesis occurs in the nucleoli themselves.

The postulate that the primary function of the nucleolus is the production of ribosomes can account for its absence during the cleavage mitoses (see Section III 2 b). Nucleoli are absent also in the nuclei of growing pollen tubes (STEFFENSEN 1966). RNA is synthesized here but its base ratios do not correspond to those of ribosomal RNA. Finally, the role of the nucleolus in ribosome formation explains why its abnormal loss leads ultimately to death.

According to VINCENT (1964) the nucleolus of the starfish appears to posses amino-acid activating and s-RNA synthesizing systems (translation components) as well as ribosomal-like particles and properties. He suggests that the transcription system is used to produce messenger-protection protein which protects "stored" messengers. There is, as we shall see, increasing evidence that such stored messengers are present in egg cells and early cleavage stages (see p. 56).

HARRIS (1965) has summarized the evidence, collected by his colleagues and himself, for the occurrence of short-lived nuclear-produced RNA molecules which are broken down within the nucleus. He is of the opinion that only a small proportion of the RNA made in the nucleus at any one time is an immediate precursor of the stable RNA found in the cytoplasm. This suggests that very much more RNA is made in the nucleus than is used as a template for the synthesis of protein. In cells which contain large amounts of nuclear "sap" the RNA in the "sap" has a bizarre and very variable base composition (Edström 1964) and varies greatly in amount from cell to cell. This is precisely what one would expect if a large part of the RNA made in the nucleus was subject to intra-nuclear breakdown. On the other

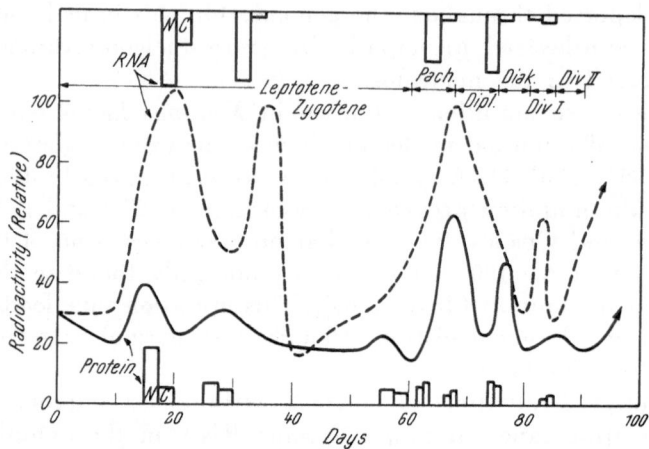

Fig. 19. Periodicity of RNA and protein synthesis in isolated microsporocytes of *Trillium erectum*. The histogram bars summarize the autoradiographic data of Taylor (1959) on lily microsporocytes, the height of the bars indicating the amount of incorporated label. N = nucleus, C = cytoplasm. (After Stern and Hotta 1963.)

hand Flamm and Birnstiel (1964) find that isolated nucleoli, but not isolated chromatin, can synthesize histone. This synthesis apparently involves an m-RNA system since it is inhibited by both puromycin and actinomycin D. Coupled with this, the decay rate of the m-RNA used for this histone synthesis is greater than for any other RNA species. Bonner (1965), therefore, suggests that the short lived RNA reported by Harris might fill this role.

b) Meiotic Cells

In many plants the pollen mother cells and grains within the anther show a striking periodicity with respect to intervals of synthesis and division. By removing microsporocytes from the anther and washing them several times Hotta and Stern (1963) were able to follow their capacity for protein and RNA synthesis. Their results are summarized in Fig. 19. Note that an interval marked by protein synthesis is also marked by RNA synthesis.

In meiotic cells of lily and tulip, RNA polymerase appears to remain with the chromosomes through the meiotic cycle (Hotta and Stern 1965). If, as this suggests, the metaphase chromosomes do contain RNA polymerase then some element in chromosome organization must prohibit its activity

during the contracted state. Prolonged pre-incubation of prophase chromosomes with mercapto-ethanol increases their template efficiency some
tenfold. Metaphase chromosomes, similarly treated, show hardly any
increase at all. Some kind of structural difference must therefore exist be-

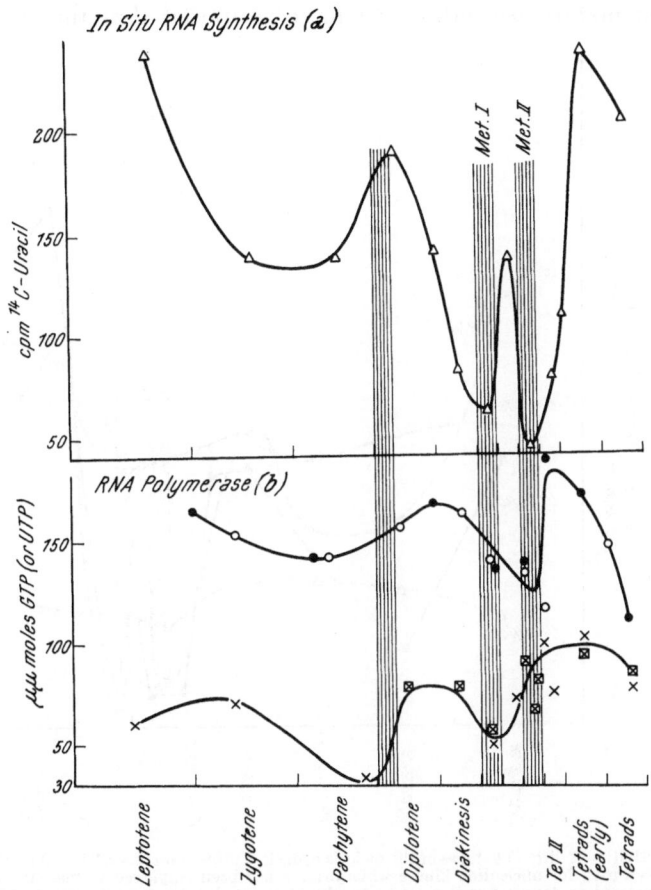

Fig. 20. RNA synthesis during the meiotic cycle in *Tulipa gesneriana*. (a) Summarizes the incorporation of C¹⁴-uracil
into microsporocyte RNA of cultured flower buds. Rates are expressed as counts per min (cpm) in purified RNA
obtained from the microsporocytes of 5 anthers. (b) indicates the RNA polymerase activity of intact nuclei and
chromosomes. Activities are expressed as moles of radioactive nucleotide incorporated into RNA in 10 min at
37° C per 100 mμ moles of DNA-P. The two studies summarized were conducted in successive years. UTP³² was
used as labelled substrate for all points marked "x", GTP³² was used for all other assays. The spacing of stages
(abscissa) corresponds roughly to natural time intervals. Stages are designated according to the preponderant cell
type which was usually about 70% of the total. The spacing between MI and II, however, represents a mixture of
AI and TI stages. (After HOTTA and STERN 1965.)

tween extended and contracted chromosomes which is capable of accounting
for such a differential response.

In an attempt to explain fluctuations in RNA synthesis, HOTTA and STERN
(1965) examined three possible mechanisms for regulating the level of synthesis in the meiotic cells of *Tulipa gesneriana*, namely variation in:

i) RNA polymerase activity,

ii) availability of template DNA, and

iii) precursor supply as determined by kinase activity.

They found that nuclear material derived from any meiotic stage had RNA polymerase activity and that such variations as were present did not parallel those for RNA synthesis *in situ* (Fig. 20). Whatever the fluctuations in polymerase activity may reflect, they cannot wholly determine the course of RNA synthesis within the meiotic cell. From the absence of RNA synthesis during metaphase one might have supposed that the enzyme would

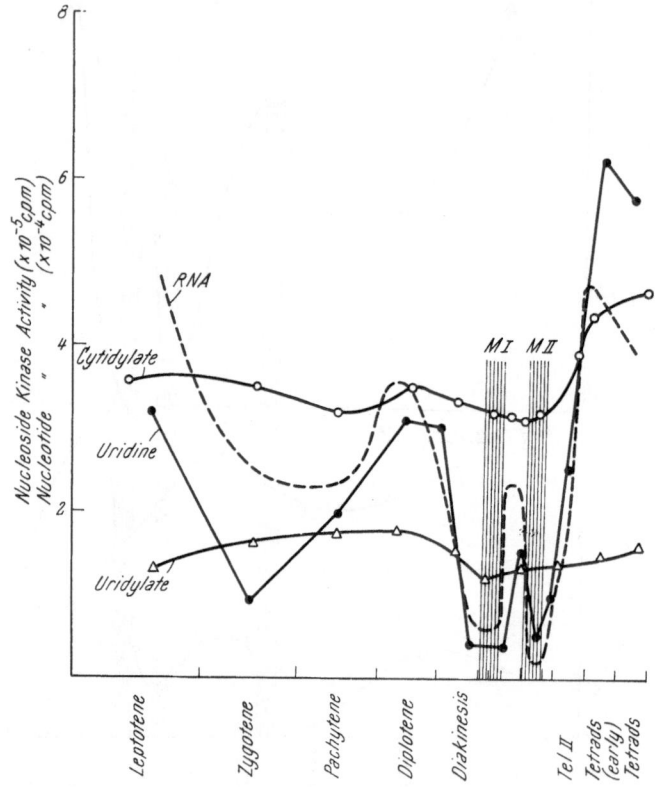

Fig. 21. Kinase activities at different stages of the meiotic cycle in *Tulipa gesneriana*. The substrates cytidylate and uridylate have been used for nucleotide kinase while uridine has been employed to measure nucleoside kinase. Activities are expressed as counts per min (cpm) in phosphorylated derivatives formed in 20 min per mg of protein in the extract. The dotted line shows the rates of RNA synthesis *in situ* during the meiotic cycle. (After HOTTA and STERN 1965.)

be shed from the chromosome at some time during prophase. In fact, however, RNA polymerase remains associated with the chromosomes throughout the cycle of contraction and extension.

The second possibility, that some control is exercised over the accessibility of enzyme to the DNA template, was also ruled out by HOTTA and STERN. On the other hand, the environment of the microsporocytes is rich in nucleic acid precursors. Since precursor supply could be limited by the kinases which control the formation of the necessary triphosphates (see p. 18), extracts of meiotic cells at dienerent stages were tested for nucleoside

Fig. 22. ³H-UdR autoradiographs of *Cyrtacanthacris tartarica* to show distribution of newly synthesized RNA at first (*a* pachytene and *b* diplotene) and second (*c*) prophase of meiosis. (Photographs kindly supplied by Dr. S. A. HENDERSON, ca. ×1 500.)

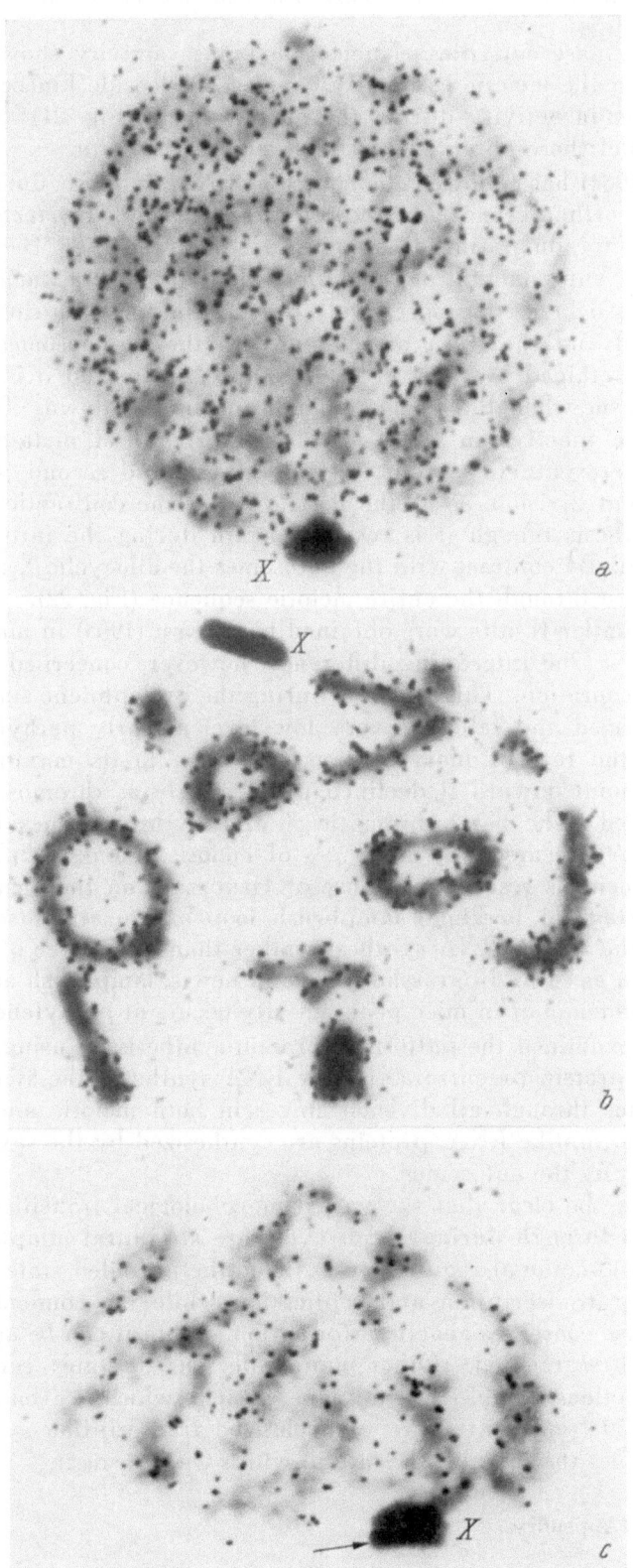

and nucleotide kinase activities. Nucleotide kinase activity showed little variation through the meiotic cycle. By contrast nucleoside kinase showed marked variation in activity during the meiotic cycle (Fig. 21) and these variations parallel those observed for RNA synthesis *in situ*.

HENDERSON (1964) has followed the course of RNA synthesis during male meiosis in the orthopterans *Schistocerca gregaria* and *Cyrtacanthacris tartarica* with autoradiographic techniques. He observed that ^{3}H-UdR was taken up by the autosomes throughout the whole of the first meiotic prophase. Synthesis decreased progressively during diakinesis as chromosome coiling increased and the RNA associated with the chromosomes at diakinesis was not retained by them at first metaphase. Instead it is released into the cytoplasm when the nuclear membrane breaks down. The autosomes which are inactive in RNA synthesis during first meta-anaphase recommence heterosynthetic activity at interkinesis and second prophase. But in the second division, as in the first, chromosome contraction signals the end of synthesis though it is resumed again during the initial stages of the spermatid. By contrast with the autosomes the allocyclic X univalent remained inactive through the entire meiotic sequence (Fig. 22).

Essentially similar results were obtained by MONESI (1965) in male meiosis of the mouse. One interesting difference, however, concerned the rate of ^{3}H-UdR incorporation. This was low during the preleptotene stage when DNA is synthesized and fell to a very low level at early pachytene. By mid-pachytene the rate of incorporation had risen to its maximal level and from this point onward it declined progressively as chromosome contraction increased. The heterochromatic X and Y chromosomes failed to incorporate RNA precursors at all stages of meiosis. MONESI explains the difference between his results and those of HENDERSON on the basis that it is the precise stage of maximal lampbrush loop organization (see p. 53) which governs the rate of RNA synthesis rather than the degree of chromosome contraction as such. In grasshoppers and newts, lampbrush activity is maximal at diplotene but in mice peak activity occurs at pachytene [*].

MONESI also examined the pattern of protein synthesis by using tritiated amino acids as protein precursors. Unlike RNA synthesis, the synthesis of protein continues through all division stages in both meiotic and mitotic cycles. Moreover, unlike RNA, proteins are synthesized by the sex chromosomes as well as by the autosomes.

It should now be clear that the varied morphological transitions which chromosomes go through during a cell cycle are structural adaptations to meet particular functional requirements. Thus the unfolded state of interphase serves for transcription and replication while the condensed state of meta-anaphase conserves genetic information so that it can be accurately distributed and segregated. Other more specialized states serve more specialized functions. Thus the polytene system which we encountered earlier (see p. 30) represents a very specialized transcription system (see p. 89). So too does the lampbrush state to which we turn next.

---

[*] See note 6 Appendix.

i) Oocytes

Throughout early embryogenesis a developing organism depends solely on the ribosomal and m-RNA synthesized during oogenesis and then conserved for future use. To this end the growing oocytes of all vertebrates and many invertebrates (insects, crustaceans, and molluscs) remain in a prolonged diplotene phase during which the bivalents increase enormously in size. Coincident with this a large number of laterally projecting loops are developed at chromomere sites from each chromatid within the bivalent. For example, in *Triturus* CALLAN has estimated that some 20,000 loops are developed.

The lampbrush chromosomes of amphibian oocytes can be readily isolated in a condition which closely resembles that in the living cell. The

Table 16. *A Comparison of the Ratios of Protein and RNA to DNA in Newt Lampbrush Chromosomes and Beef Liver Chromatin.*
(Data of IZAWA et al. 1963.)

Item	Protein/DNA	RNA/DNA	DNA per Diploid Set
1. Beef liver chromatin	2.8	0.062	—
2. Newt lampbrush chromosomes	550	9.0	0.39×10^{-9} gm
3. Newt erythrocyte chromosomes	—	—	0.089×10^{-9} gm

loop axes and interchromomeric thread of the main axis of isolated chromosomes can be broken by DNase (MACGREGOR and CALLAN 1962) but not by protease or RNase. The axes of the lateral loops, on the other hand, are coated with an RNP (ribonucleoprotein) matrix which can be removed with proteases or RNase but is not affected by DNase. Labelling with ^{3}H-uridine shows that the overwhelming majority of lateral loops take up the radio-isotope uniformly throughout their lengths (GALL and CALLAN 1962). Inhibitors of RNA synthesis, such as calf thymus histones (see p. 65) and actinomycin D cause the collapse of the loop system (IZAWA et al. 1963).

Compared with the genetic apparatus of an active, differentiated cell the lampbrush chromosome is a site of massive gene activity. Thus the ratios of both RNA and protein to DNA are very much higher than in somatic nuclei and somatic chromosomes (Table 16).

Like polytene chromosomes, lampbrush chromosomes are thus synthetically active at a time when the chromosomes are visible. There are, however, two interesting differences between the two systems. First, while loops occur at almost all the chromomeres of lampbrush chromosomes, puffs are found at only about 2% of the bands of polytene chromosomes (BEERMANN 1952). Second, lampbrush chromosomes subsequently give rise to regular chromosomes which have a future in heredity. Eventually all the lateral loops regress and the chromosomes pass into diakinesis. By contrast polytene chromosomes do not normally return to a regular state. Nor do they ever again divide. Despite these differences the two states have much in common. The puffed state of the bands in polytene nuclei is in many ways comparable to the lampbrush condition of the chromomeres in meiotic cells. Heterosynthetic activity in the polytene nuclei of somatic cells, however, generally differs from that in oocytes in being highly localized, only a small number of bands being in a puffed state at any one time in any one tissue. Recent investigations by BIER, KUNZ, and RIBBERT (1968) have shown that this difference reflects the nature and function of the cell rather than the state of the chromosomes.

These workers compared the pattern of oogenesis and RNA synthesis is insects with two types of ovary, namely the panoistic type in which the oocytes are surrounded only by a follicular epithelium and the meroistic type where, in addition to the epithelium, each oocyte is connected with one or more nurse cells until the end of the growth period.

Meiosis in panoistic ovaries is characterized by the presence of lampbrush chromosomes comparable with those in amphibia. Multiple nucleoli are found also and RNA synthesis is pronounced in both the lampbrush chromosomes and the nucleoli. The situation is quite different in insects with meroistic ovaries. Thus in *Musca domestica* and *Calliphora erythrocephala* all the oocyte chromosomes condense and aggregate to form a single compact karyosome and they barely participate in RNA synthesis. Even so, the oocytes of meroistic ovaries grow more rapidly than those in panoistic types and they show high rates of cytoplasmic protein synthesis. Autoradiographic studies have shown that this synthesis is sustained by RNA produced by the nurse cells which is passed into the ooplasm via cytoplasmic bridges.

Thus, the synthetic functions performed by the lampbrush chromosomes in panoistic ovaries are taken over by the nurse cell nuclei in meroistic ovaries. In *Calliphora*, the nurse cell nuclei contain banded polytene chromosomes but the pattern of RNA synthesis in them differs from that in, for example, the polytene nucleus of the trichogene cell of the bristle-forming apparatus. Thus, both short and long treatments with ^{3}H-UdR result in more or less uniform labelling along the length of the chromosomes in nurse cells. Further, irregularly-shaped, nucleolar-like aggregations of ribonucleoprotein which show intense incorporation of RNA precursors are found at several different loci. Thus, although the structure of the nurse cell nucleus is comparable with that of polytene nuclei in non-germinal organs, its pattern of synthesis is more similar to that of the lampbrush chromosomes in oocytes.

The quantity of DNA per set of chromosomes is often constant for the different cells of a given organism (BOIVIN et al. 1948, MIRSKY and RIS 1949). Bearing in mind that the DNA level of an erythrocyte nucleus is 2C while that of an oocyte is 4C, the DNA content per chromosome set of the lampbrush system is in fact some four times that of the erythrocyte chromosomes in the same organism (Table 16). It would appear that increased RNA synthesis in the lampbrush system is based on an augmented number of DNA templates.

By direct measurement of the amount of DNA which is complementary to newly synthesized RNA in stage-4 (maximum lampbrush activity) oocytes it is possible to obtain an estimate of how much of the genome is active in lampbrush synthesis. Thus hybridization experiments between ^{3}H-DNA, extracted from the testis, and ^{32}P-labelled, stage-4 oocyte RNA of *Xenopus laevis* show that only about 1.5% of the DNA is complementary to lampbrush RNA (Fig. 23). Since only one strand of the DNA duplex is expected to be active in RNA synthesis *in vivo* this implies that some 3% of the total genome is functional at this time (DAVIDSON et al. 1966). The lampbrush chromosome is thus distinctive not in the proportion of its genome which is active but rather in its role as a system for elaborating large quantities of extremely long-lived gene products. Thus the astonishingly high protein content of the chromosomes and the intense protein synthesis occurring continuously on the lampbrush loops (Table 17) both suggest that the newly synthesized RNA is packaged *in situ* within newly formed protector proteins.

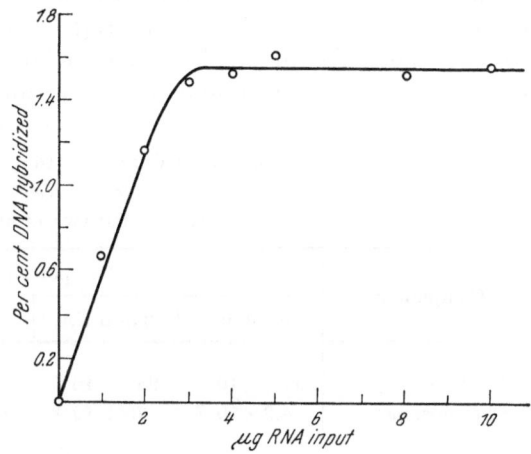

Fig. 23. Typical saturation curve for the formation of ribonuclease resistant hybrids between testis H[3]-DNA and stage 4 (maximum activity) lampbrush P[32]-RNA of *Xenopus laevis*. Each 200 ml of annealing mixture contains 5 µg DNA and the amounts of RNA plotted on the abscissa. The specific activity of the RNA was 914 cpm/µg. (After DAVIDSON et al. 1966.)

Now the oocyte nucleus contains twice as much DNA as is found in the chromosomes (Table 17). Clearly there is as much DNA in the nucleoplasm as in the chromosomes themselves. An essential feature of the lampbrush oocyte is the immense number of "nucleoli" formed. In *Triturus*, for example, the number has been estimated to be in excess of 1,000 (GALL 1968) while for *Xenopus* the comparable figure is 600–1,200. GALL (1968) has recently demonstrated that in this case the formation of nucleoli is governed by a differential replication of the nucleolar organizer region during pachytene. This can be followed autoradiographically using ^{3}H-TdR. With this treatment the nucleolar granules take up tritium whereas the chromosomes remain unlabelled. The extra organizers produced migrate to the nuclear envelope where they serve as templates for the synthesis of r-RNA precursors (MACGREGOR 1967). The extra DNA, formed by amplification of the organizer region, is metabolically stable and persists throughout oogenesis in association with the nucleoli. This leads to large increases in the DNA content of the oocyte nucleus. Thus in *Xenopus* the DNA content rises from 4 to 14 C while in *Bufo* the value attained is in the region of 10 C. Two general mechanisms appear to lead to such a multinucleolate condition (MACGREGOR 1965). In some species all peripheral nucleoli are formed before or soon after the chromosomes assume a lampbrush form. This occurs, for example, in *Triturus cristatus carnifex* and it implies that no part of the lampbrush system is involved in the process which adds to the nucleolar population. In other species (*e.g. T. c. cristatus* and *Plethodon cinereus*) many of the peripheral nucleoli are produced by, and shed from, one or more nucleolar organizers on the lampbrush chromosomes. In plethodont salamanders, for example, the nucleoli are represented by 200–300 ring-shaped structures which vary greatly in size. These rings give a positive Feulgen reaction which indicates that they have a DNA component. Significantly DNase disrupts the rings. RNase and protease do not disrupt them but these enzymes do remove nucleolar material. These "nucleoli" thus appear to be circular molecules of DNA with associated RNP complexes (MILLER 1965). In fact these rings originate from the nucleolar organizer which remains

Table 17. *DNA, RNA and Protein Content of Oocyte Nucleus and Lampbrush Chromosomes of the Newt.*
(Data of IZAWA et al. 1963.)

Component	Gm per Nucleus				Pr/ DNA	RNA/ DNA
	Protein	Total NA	DNA	RNA		
1. Nucleus	17×10^{-7}	28×10^{-9}	1.6×10^{-9}	26.4×10^{-9}	1060	16.5
2. Chromosomes	4.3×10^{-7}	7.8×10^{-9}	0.78×10^{-9}	7.0×10^{-9}	550	9.0

in an extended state long after other loops have retracted. One further difference is that the nucleolar loop, unlike the others, does not arise as a paired structure. A filamentous component thought to represent loops analogous to those formed on lampbrush chromosomes has also been found in plant nucleoli following treatment with Tween 80 (LA COUR and WELLS 1967).

These observations imply that the nucleoli of lampbrush oocytes represent a device for the amplification of those DNA units which code predominantly for ribosomal RNA—a system of highly selective gene replication. This clarifies the biochemical finding that while quantitatively one would expect the loci responsible for ribosomal RNA synthesis to amount to only < 0.1–0.3% of the total genome (YANOFKSY and SPIEGELMANN 1963) yet per unit of time some 97–99.5% of the RNA synthesized in lampbrush oocytes is ribosomal in character (DAVIDSON and MIRSKY 1965).

A comparable escalation of nucleolar organizing DNA can be demonstrated in other oocytes too. Thus in the oogonial interphase of tipulid flies a DNA body is present which at each mitosis is included in only one of the two anaphase groups (BAUER 1932, BAYREUTHER 1952). The DNA of this body, like that of the chromosomes themselves, is complexed with histone. The body increases appreciably during the pre-meiotic interphase and, in the oocytes, nucleoli are found inside it (LIMA-DE-FARIA 1962, LIMA-DE-FARIA and MOSES 1966). Since the body contains about 59% of the total DNA of the nucleus, it has been suggested that this too represents many multiples of the nucleolar organizing region. Similar structures exist also in the oocytes of the beetle *Dytiscus marginalis* (URBANI and RUSSO-CAIO 1964) * and in the cricket *Acheta domestica* (NILSSON 1966, LIMA-DE-FARIA et al. 1968). Here the DNA body appears first in oogonia at interphase. It synthesizes DNA at a different time from the chromosomes themselves and persists through all the pre-meiotic mitoses. During pre-meiotic interphase and early first meiotic prophase, the body increases considerably in size. Then at pachytene-diplotene an outer shell of RNA forms around the inner core of DNA and histone. Towards the end of diplotene the structure starts to disintegrate, releasing DNA, histone and RNA into the nucleoplasm.

To what extent a comparable augmentation of DNA occurs at active loci of interphase chromosomes is not known. It is worth bearing in mind that if the percentage of active loci in such nuclei is the same as the percentage of puffs in polytene systems then the increase in DNA content would hardly amount to a significant departure from "constancy".

Ovulated, unfertilized eggs of *Rana pipiens* and *Xenopus laevis* contain some 300–500 times the diploid quantity of DNA. Hybridization experiments show this DNA to be complementary with only a small fraction of liver DNA from the same species. Indeed this DNA is not represented at all in erythrocytes (DAWID 1965). DAWID (1966) has shown that the bulk of this egg DNA is of mitochondrial origin and depends upon the large size of the egg cell and the very large number of mitochondria per cell. During cleavage these mitochondria are partitioned, without change, into the newly formed cells. Their behaviour is thus analogous to that of the egg ribosomes (see above) and can be related to the fact that the oocyte has to supply both ribosomes and mitochondria for a mass of cytoplasm equivalent in quantity to several thousand ordinary somatic cells. Thus, whereas a somatic cell on average contains some 3.3×10^6 ($= 12$ picogm.) of ribosomal RNA, a mature oocyte of *Xenopus* has 1.1×10^{12} ($= 4 \mu$gm.) of it (PERKOWSKA et al. 1968).

ii) Spermatocytes

A second instance of the modification of chromosome structure at sites of gene activity is found in growing spermatocyte nuclei of species of *Drosophila*. This reaches its most differentiated form in the *D. hydei* sub-group (HESS 1967). Here the Y-chromosome may be considered as a univalent lampbrush chromosome with a few (5) very large loops. Each pair of loops can be identified by its specific morphological characteristics (Fig. 24). These loops are active in the production of RNA used in producing proteins needed during spermatogenesis. Thus in XO males spermatogenesis is blocked at the spermatocyte stage. If the clubs and tubular ribbons are missing development is blocked at the early spermatid stage. When threads and pseudo-nuclei are missing, undersized immobile sperm result.

The small number and enormous size of the *Drosophila* loops suggest that they exist for amplifying informational units necessary for the production of the structural proteins which are required in large amounts for the formation of the giant sperm found in the genus. Since only one-half

---

* See note 7 Appendix.

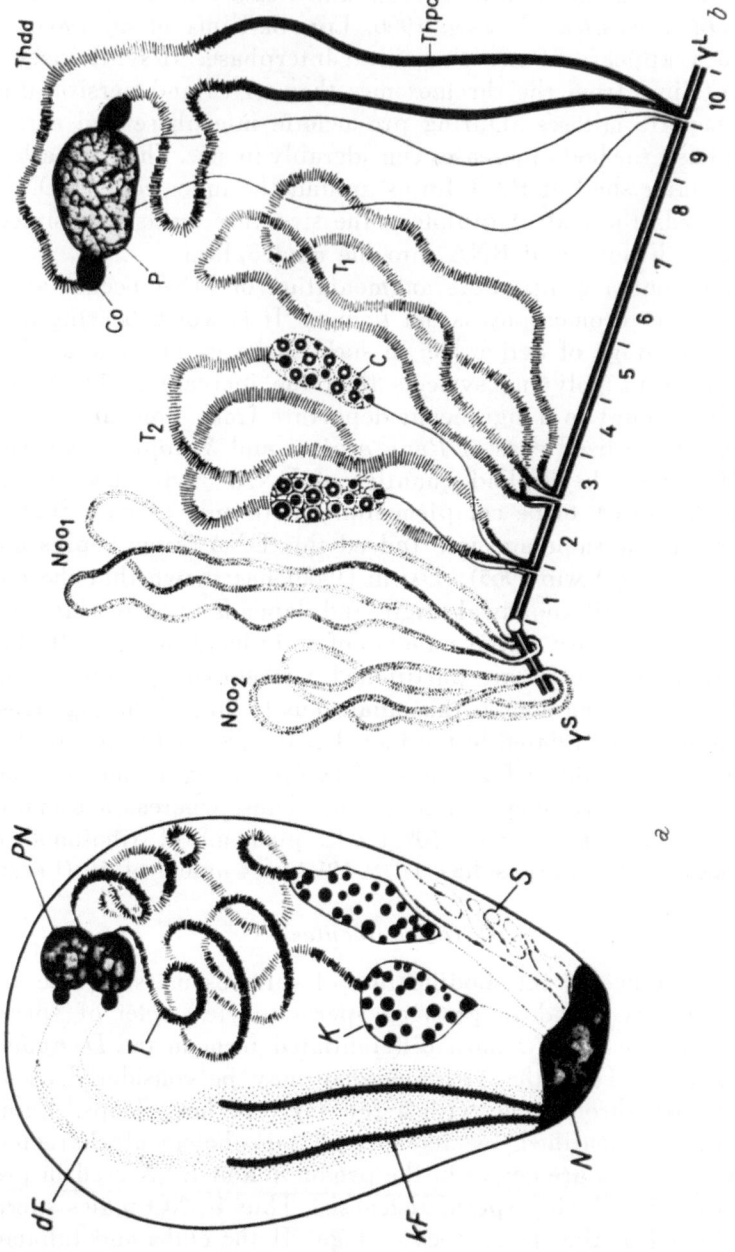

Fig. 24. Diagrammatic representation of the lampbrush activity of the Y-chromosome in *Drosophila hydei*. *a* (after Hennig 1967) shows the intact spermatocyte nucleus, while in *b* (after Hess 1967) the patterns and positions of the individual loops in the long arm of the univalent Y are indicated. C = K = Clubs, C_0 = cones of pseudonucleus, Noo_1 and Noo_2 = S = nooses, P = PN = pseudonucleus, T_1, T_2 = T = tubular regions, Thdd = dF = distal diffuse section of the threads, Thpc = KF = proximal compact section of the threads.

of the meiotic products contain a Y these proteins must be made before the spermatids are formed so that they too may be equally partitioned during the meiotic process.

The earlier studies on the lampbrush-Y system do not clearly demarcate the precise stage at which the Y-loops form, it being referred to simply as the "growth phase of the spermatocyte" (see for example Meyer 1963, Hess

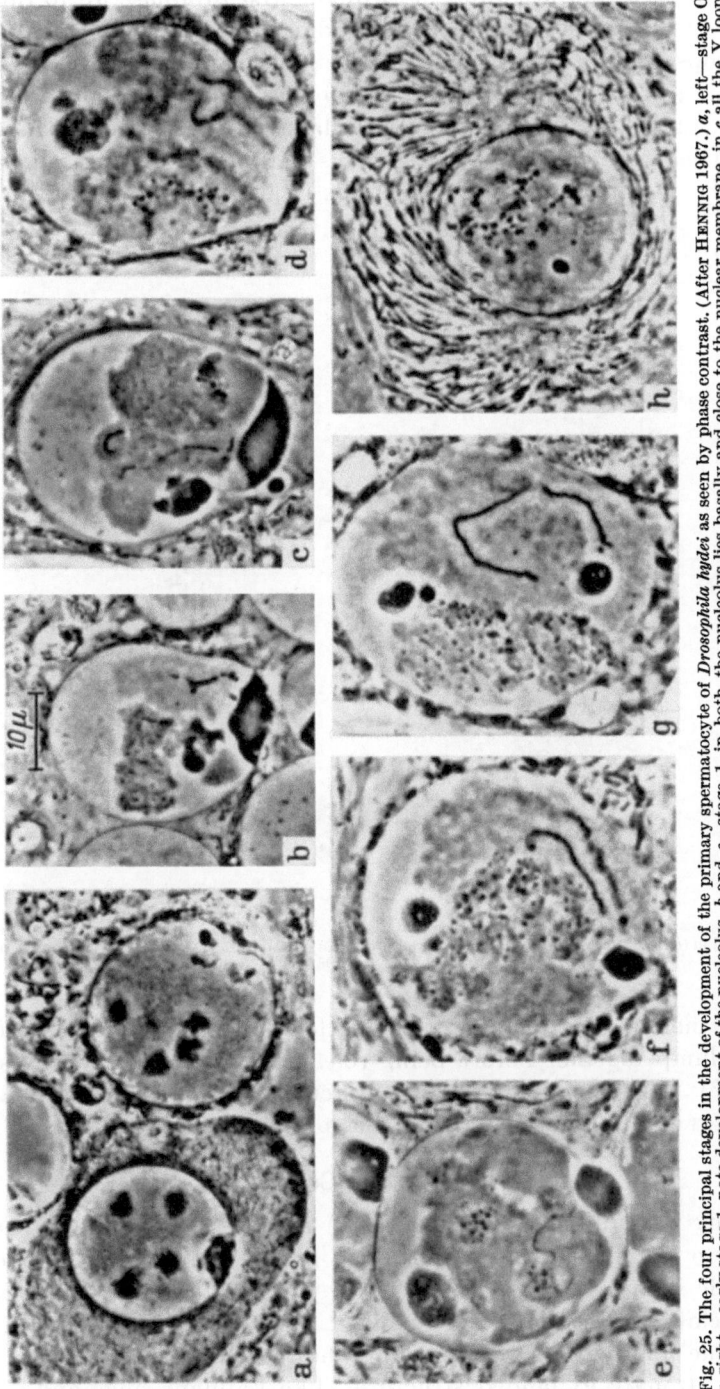

Fig. 25. The four principal stages in the development of the primary spermatocyte of *Drosophila hydei* as seen by phase contrast. (After HENNIG 1967.) *a*, left—stage 0, *a*, right—early stage 1, note development of the nucleolus, *b* and *c*—stage 1, in both the nucleolus lies basally and close to the nuclear membrane, in *c* all the Y loops are recognizable, *e*—late stage 2, the nucleolus has become much smaller and is now detached from the nuclear membrane, *f*—stage 3, the diffuse threads are now not so voluminous as in the earlier stages and the clubs are beginning to show their characteristic form, *g*—onset of stage 4, the Y products are beginning to disappear, *h*—stage 4 (= diakinesis) the spindle is already forming and the Y products are still further dispersed.

1967). The recent work of Hennig (1967) shows that lampbrush activity precedes diakinesis and occupies most of the first prophase up to this point (Fig. 25). Incorporation studies with radioactive precursors have shown that the loops contain axial DNA and actively synthesize RNA. Labelled RNA is stored in the loops for about 20–30 hours while the loops themselves persist for 120 hours. Incubation with labelled amino acids shows that labelled nuclear proteins leave the nucleus within a few hours and are not bound to the Y-loops on a long-term basis (Hennig 1967). In late prophase

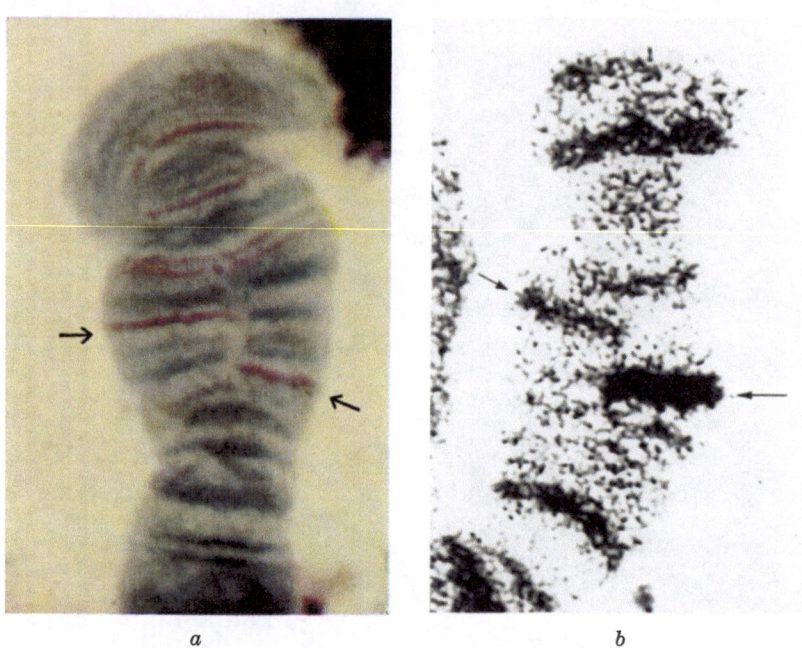

a   b

Fig. 26. Homologous regions of section 14 in chromosome III of *Chironomus tentans* as seen in a toluidine blue stained preparation (*a*) and an autoradiograph (*b*) taken after 6 hr incubation in ³H-UdR. The arrows indicate the puff 14 C 2. (After Pelling 1964, ca. × 1,360.)

of the first meiotic division the Y begins to condense and the products of the loops clump together before being released into the developing spindle (Meyer 1963). These remnants persist intact until after first metaphase and in some cases can be traced as far as the spermatid stage.

c) Polytene Cells

Following the studies of Rudkin (1963) and Pelling (1964) is has been customary to argue that the majority of bands in polytene chromosomes contain little if any RNA. It is of course possible that RNA is present in amounts below the level of detectability with ³H-UdR incorporation or toluidine blue staining. There is no doubt, however, that the average amount of RNA in strongly expanded structures, like Balbiani rings and strongly puffed bands, is greater than that of weakly puffed or non-puffed bands (Fig. 26). Edström and Beermann (1962) likewise observed that the puffs have a higher RNA/DNA ratio than other parts of the giant chromosomes (Table 18). For example in chromosome IV the second Balbiani ring (BR II) contains some 20 $\mu\mu$g of RNA so that the RNA/DNA quotient of this giant puff is some 400 times higher than the average for chromosome I.

Table 18. *Distribution of RNA/DNA in the Chromosomes of Chironomus tentans.*
(After EDSTRÖM and BEERMANN 1962.)

Component	RNA/DNA
1. Whole chromosome set	1/7–1/8
2. Chromosome I	1/20
3. Chromosome IV	
(a) Mid-segment	1/3
(b) Balbiani ring (BRII)	20/1

Direct analysis of the base composition of the RNA from different chromosomes and different chromosome regions is possible by microelectrophoresis. This technique shows that chromosomal RNA is distinguishable from both nucleolar and cytoplasmic RNA (Table 19). It is also shows that

Table 19. *Base Composition in Percent of RNA Obtained from Different Regions of the Polytene Nuclei of Chironomus tentans.*
(After EDSTRÖM and BEERMANN 1962, EDSTRÖM 1964.)

		Base Composition			
		A	G	C	U
Nucleolus	Chromosome II	31.0 ± 0.3	22.8 ± 0.2	18.5 ± 0.3	27.7 ± 0.4
	Chromosome III	30.8 ± 0.4	22.2 ± 0.3	18.8 ± 0.6	28.3 ± 0.5
Cytoplasm		29.4 ± 0.4	22.9 ± 0.3	22.1 ± 0.4	25.7 ± 0.3
Chromosome IV	Balbiani ring 1	35.7 ± 0.6	20.6 ± 1.7	23.2 ± 1.2	20.8 ± 0.8
	Balbiani ring 2	38.0 ± 0.6	20.5 ± 0.6	24.5 ± 0.6	17.1 ± 0.6
	Balbiani ring 3	31.2 ± 2.2	22.0 ± 2.0	26.4 ± 1.9	20.2 ± 1.4
	Average for chromosome IV	36	21	24	19
Average for chromosome I		29.4 ± 0.5	19.8 ± 1.0	27.7 ± 0.8	23.1 ± 0.6

the precise composition of chromosomal RNA varies from segment to segment within the polytene system. For example the RNA derived from the Balbiani rings is notable for its very high adenine content. Indeed puff RNA with its unique base asymmetry has at least some of the characteristics expected of m-RNA (BEERMANN 1967). In *Chironomus tentans*, puffs and Balbiani rings contain a basic protein which differs from chromosomal proteins in most respects but which shows similarities with the basic structural protein of ribosomes (Lezzi 1967). This finding is in agreement with the hypothesis that ribosomes participate in the build up of puffs and Balbiani rings.

In *Sciara coprophila* a large number of sites, approximately 18% per chromosome (Table 20), give rise to micronucleoli (Fig. 27). The majority of these micronucleoli form at condensed bands but they are found also at DNA puffs (Gabrusewycz-Garcia and Kleinfeld 1966).

Table 20. *Distribution of Micronucleolar forming Sites in Sciara coprophila.*
(Based on Gabrusewycz-Garcia and Kleinfeld 1966.)

Chromosome No.	Total No. of Bands	No. of Micronuclear Sites
1. II	240–250	43–50
2. III	290–300	52–55
3. IV	340–350	68–70

d) Endoploid Cells

Prescott (1964, 1966) has followed the synthesis of RNA and protein in the macronuclei of *Euplotes*. Here, it will be recalled (see p. 42), the cell cycle has no measurable G_2 period while during S the synthesis of DNA is restricted to two narrow replication bands. RNA synthesis takes place during G_1, S and continues during amitotic splitting. The only time it is not detectable is during, and at the site of, DNA replication, i.e. in the replication bands. Electron microscope examination of the *Euplotes* macronucleus shows that the presence and absence of RNA synthesis is related to the organization of the chromatin (Kluss 1962). Nuclear protein synthesis too takes place throughout interphase (Fig. 28), though, as Gall (1959) has shown, histone synthesis is confined to the site of DNA synthesis in the replication band. These findings thus support the conclusions that:

(i) DNA and histone synthesis are closely associated events, and

(ii) DNA engaged in replication is unable to support RNA synthesis.

The hypopharyngeal glands, located in the head of the honey bee, are responsible for forming the royal jelly—a mixture of proteins, lipids and vitamins. This is synthesized and secreted during a short period of the adult workers life (days 5–11). The cells of the glands begin to differentiate, however, in very young pupae. As they differentiate they undergo endomitosis (Painter and Biesele 1966). This endomitotic cycle involves three characteristic phases:

(i) an interphase which is characterized by a large number of nucleoli,

(ii) a prophase-like condition in which the nucleoli fragment releasing both single ribosomes and large quantities of nuclear synthesized protein into the nuclear sap; the ribosomes separate from the proteins and enter the nuclear pores where they are organized into polysomes, and

(iii) a late prophase state in which the chromosomes appear in bundles and all nucleoli have disappeared.

A new series of nucleoli appear in the next interphase period and the

Additional material from *The Chromosomes Cycle,*
ISBN 978-3-663-12615-7, is available at http://extras.springer.com

whole process is repeated again. PAINTER (1966) sees these endomitotic cycles as a means of boosting the polysome forming capacity of the hypopharyngeal cells in preparation for its subsequent synthetic role.

In the Cecidomyidae from 1–23 E-chromosomes are eliminated from the soma during the first half-dozen cleavage divisions but are retained in the germ line. Recently PAINTER (1966) has suggested that the presence of

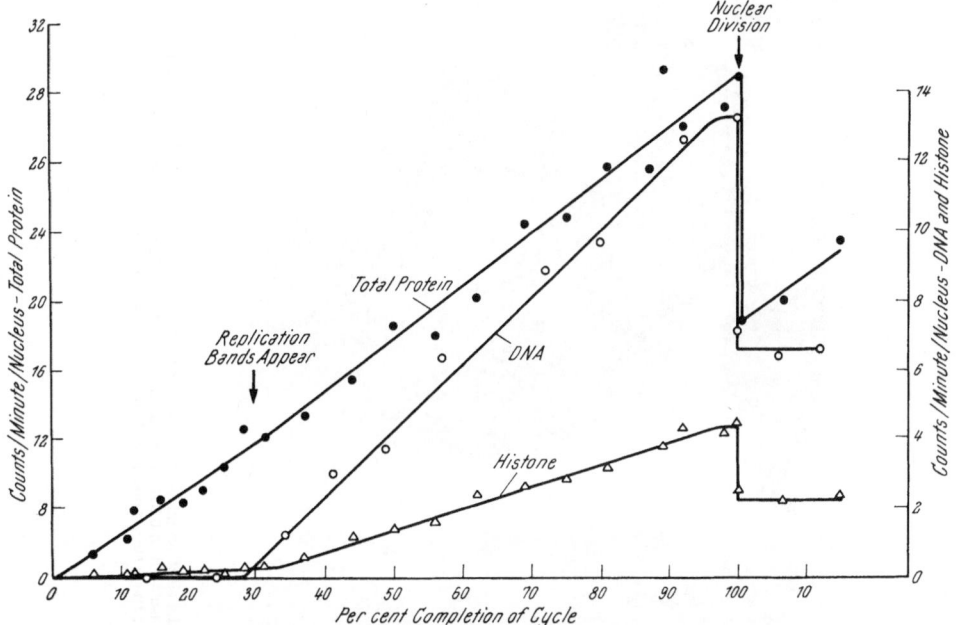

Fig. 28. Synthetic activity in the macronucleus of *Euplotes eurystomus* during the cell cycle. Each point is the mean count of 30–60 nuclei following the incorporation of ³H-TdR (DNA) and tritiated amino-acids (Total protein and histone). Notice there is no G₂ period in the macronuclear cycle. (After PRESCOTT 1966.)

such E-chromosomes serves "to increase the polysome forming capacity of the nurse cells in egg follicles" and so accomplishes the same end as a series of endomitotic cycles. In a like manner, he suggests that the retention of the heterochromatic, acentric, end segments in the polycentric germ-line chromosomes of *Parascaris equorum* serves to increase the polysome capacity of the germ cells, especially during egg formation.

e) Non Mitotic Diploid Cells

The nuclei of calf thymus lymphocytes can survive isolation in isotonic sucrose solution containing a small amount of calcium chloride. Nuclei so prepared retain their soluble proteins and nucleoproteins, including soluble enzymes (STERN and MIRSKY 1953) and are capable of at least three classes of DNA dependent synthetic reactions:

(a) the synthesis of ATP,

(b) the incorporation of amino acids into nuclear proteins, and

(c) the incorporation of many purine and pyrimidine precursors into RNA and DNA.

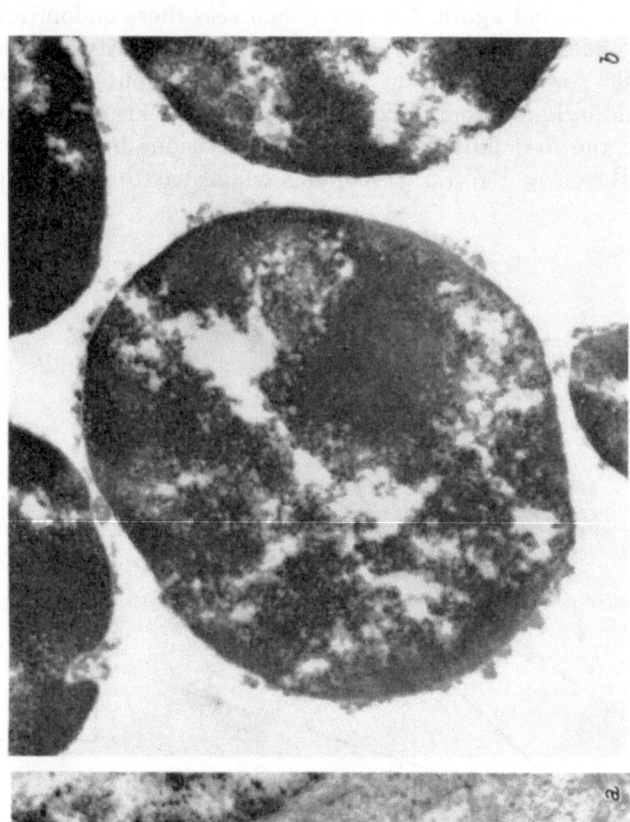

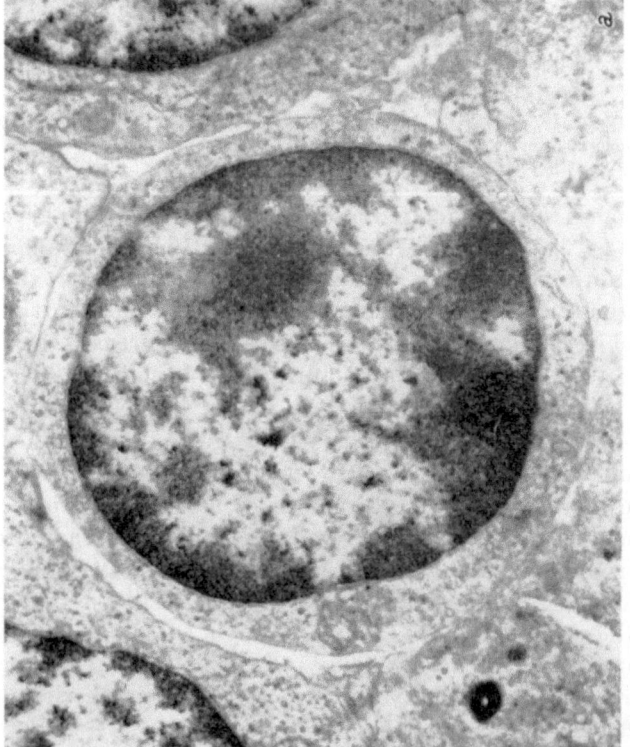

Fig. 29. Structure of the calf thymus lymphocyte interphase nucleus as seen with the electron microscope. (a) is an intact cell (ca. ×15,000) following fixation in 1% OsSO₄ pH 7.0 in acetate-veronal buffer at 0° C while (b) is a nucleus (ca. ×20,000) which has been isolated in 0.01M citric acid and suspended in a sucrose-phosphate medium. In both notice that the chromatin occurs in both a dense and a diffuse condition though the latter is not as well preserved as in the intact cell. (After LITTAU et al. 1965.)

In these lymphocytes the greater part of the chromatin is arranged in dense masses and only in relatively small areas is the chromatin diffuse (Fig. 29). The chromatin masses are composed of a dense reticulum of 100 Å fibrils and are similar to those reported for nucleated erythrocytes (see p. 74) and chromocenters (FRENSTER, ALLFREY, and MIRSKY 1963). When isolated nuclei are incubated with radioactive RNA precursors the newly synthesized RNA is found almost exclusively within the extended microfibrils of the diffuse chromatin. Indeed metabolic studies reveal that some 80% of the DNA in these nuclei is inactive in m-RNA synthesis and must therefore exist in a repressed state. Both the condensed and the extended chromatin can be isolated under gentle conditions. The differential morphology and activity of this isolated material parallels that found in the

Table 21. *Base Composition of DNA Isolated from Interphase Chromatin Fractions of Calf Thymus Lymphocytes.*
(After FRENSTER 1965.)

Chromatin Fraction	Moles per 100 Moles of Total Bases Present as			
	A	T	G	C
1. Condensed	27.9	27.7	22.85	21.55
2. Diffuse	29.2	27.8	22.15	20.85

intact nucleus. No significant differences are found in the average base composition of the two chromatin fractions (Table 21). Likewise, following extraction, the total histone content of the two fractions relative to that of DNA is not significantly different and within each fraction some 20% of the total histone is of the lysine-rich variety *. By contrast diffuse chromatin contains a two-fold excess of total non-histone proteins, a five-fold excess of total RNA and of total phospholipids and an almost four-fold excess of total phosphoprotein phosphorus (Table 22).

LITTAU et al. (1965) have presented evidence that histone holds the chromatin in clumps in the dense areas of the nucleus. By removing histones selectively from isolated thymus nuclei and examining the nuclei with the electron microscope after the extraction they found that:

(i) Only removal of the lysine-rich histones, which comprise some 20% of the total histone, caused the condensed chromatin to dissociate into a diffuse network of fibrils. The removal of the arginine-rich histones had no such effect.

(ii) By restoring lysine-rich histones to histone-depleted nuclei it was possible to restore the dense chromatin masses (Fig. 30). Again arginine-rich histones had no such effect. Indeed if a mixture of the two histones was added to the histone-depleted nuclei the presence of the arginine-rich histones prevented the lysine-rich histones from clumping the chromatin threads.

Evidently, though both histone types are combined with the phosphoric acid groups of DNA, there is a fundamental difference in the manner of

* See note 8 Appendix.

their combination in dense and diffuse chromatin. Littau et al. have argued that lysine-rich histone molecules in dense chromatin cross-link the DNA fibrils whereas arginine-rich histones combine alongside them. It is this cross-linking of fibrils that leads to the production of a relatively impenetrable system of fibrils which characterized dense chromatin. But why lysine-rich histone is capable of such extensive cross-linking under some conditions though not others remains unresolved *.

Histone proteins extracted from lymphocyte nuclei are capable of repressing the template function of DNA in a variety of cell-free systems. Such repression appears to be mediated by the ability of these histones, as

Table 22. *Chemical Relationship of Nuclear Constituents within Isolated Chromatin Fractions of Calf Thymus Lymphocytes.*
(After Frenster 1965.)

Item	Mg per 100 mg DNA	
	Diffuse Chromatin	Condensed Chromatin
1. Total RNA	9.0 ± 4.7	1.8 ± 0.7
2. Total phospholipids	17.0 ± 3.6	3.7 ± 1.3
3. Total histones	90.7 ± 7.7	101.0 ± 9.1
4. Non-histone residual protein	109.0 ± 6.1	54.9 ± 5.7
	Moles P per 100 mg DNA	
5. Total phosphoproteins	13.5 ± 0.9	3.79 ± 0.6

cationic polyelectrolytes, to form electrostatic complexes with DNA by interacting with the negative phosphate groups of polyanionic DNA. This precludes the separation of the individual polynucleotide strands of the DNA dimer and so prevents the transcription which leads to m-RNA synthesis. On the other hand, when the synthetic polyanion polyethylene sulphonate is added to incubations of isolated active (diffuse) and repressed (dense) chromatin a marked increase in m-RNA synthesis occurs in each fraction. Various natural nuclear polyanions have a similar ability to de-repress the synthesis of RNA in isolated repressed chromatin fractions but little or no effect is observed on the rate of RNA synthesis in active chromatin. Indeed each of the nuclear polyanions is capable of de-repressing RNA synthesis when added to incubations of repressed chromatin (Fig. 31). The resistance of active chromatin to such additions implies that the de-repression mechanism within active chromatin is already fully saturated by natural nuclear polyanions and can only respond to excessively strong synthetic polyanions such as polyethylene sulphonate.

Frenster (1965 a, b) has therefore proposed that the DNA within repressed chromatin has most of its negative phosphate groups neutralized by polycationic histones which allows such neutralized DNA to assume a highly condensed state (see also p. 79). By contrast fewer of the negative

---

* See note 9 Appendix.

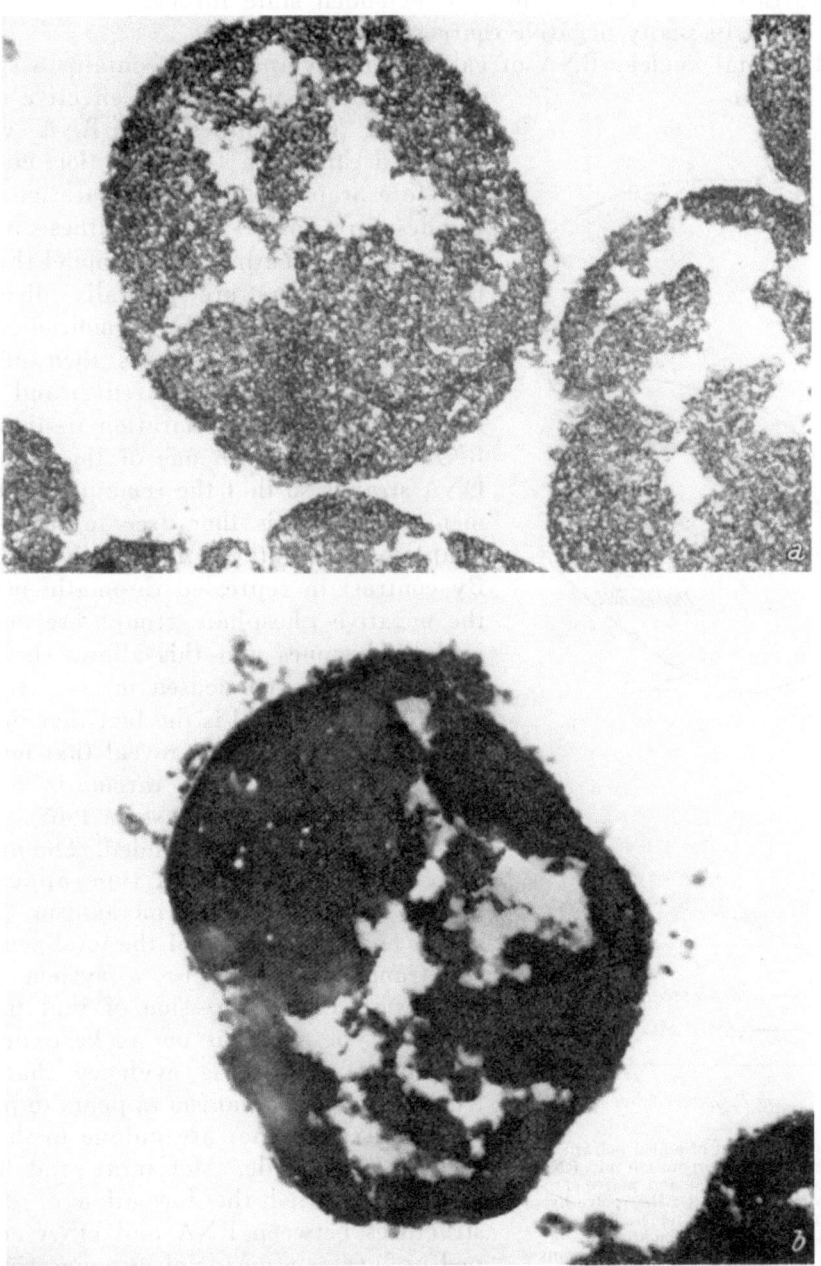

Fig. 30. The effect of removing lysine-rich histone from isolated calf thymus lymphocyte nuclei (*a*). Notice that the chromatin now has the appearance of a network of fibrils. (*b*) is a nucleus from which lysine-rich histone has been extracted and then twice the amount of the lysine-rich fraction removed has been re-added. Note restoration of strongly condensed chromatin. (After LITTAU et al. 1965, ca. × 20,000.)

phosphate groups in the DNA of active chromatin are neutralized since within such chromatin many of the polycationic histones are partially displaced from DNA by the action of nuclear polyanions. The DNA within active chromatin thus assumes an extended state through the mutual repulsion of its many negative charges.

The total nuclear RNA of calf thymus lymphocytes contains a species of RNA that is particularly effective in de-repressing the synthesis of RNA within repressed chromatin. Frenster (loc. cit.) has therefore proposed a strand-separation model for de-repression of RNA synthesis within interphase chromatin. In this model (Fig. 32) histone repressors are partially displaced from a portion of the DNA molecule. The segment freed of histone is then able to undergo intermittent localized strand separation. During this separation de-repressor RNA hybridizes with one of the separated DNA strands so that the remaining complementary strand is then free to serve as a template for synthesis of specific m-RNA. By contrast in repressed chromatin most of the negative phosphate groups are neutralized by histones and this allows the DNA to collapse into condensed masses. In support of such a model is the fact that thermal hyperchromicity studies reveal that much of the DNA within active chromatin is in a single-stranded state (Frenster 1965).

This model can be extended. The nucleus of each differentiated cell type appears to possess an epigenetic mechanism which selects specific portions of the total genotype for transcription or else a system which ensures selective repression of that part of the genotype which is not to be expressed. Significantly, there is evidence that each tissue within an organism appears to possess species of RNA that are unique to that tissue. For example, McCarthy and Hoyer (1964) have used the formation of duplex structures between RNA and DNA entrapped in agar as a means of assessing similarities and differences among RNA populations in different tissues of the mouse. With this technique large differences were demonstrated among the rapidly labelled RNA molecules isolated from different organs.

Chromatin isolated from various tissues of the pea plant will support DNA-dependent RNA synthesis in the presence of the RNA polymerase of E. coli. By coupling this chromosomal RNA-generating system to an

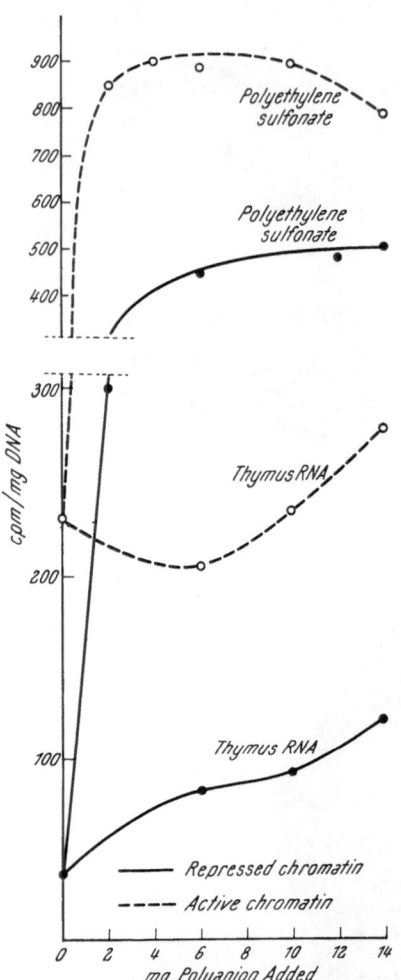

Fig. 31. The effect of added polyanions on UTP-2-C[14] incorporation into RNA of isolated repressed and active chromatin fractions. Notice that polyethylene sulphonate can de-repress RNA synthesis within both chromatin fractions while thymus RNA de-represses only repressed chromatin. (After Frenster 1965 b.)

m-RNA dependent ribosomal protein synthesizing system also derived from *E. coli* it is possible to study the production of the reserve globulin which is normally synthesized *in vivo* only in developing pea cotyledons. Under the direction of chromatin isolated from developing pea cotyledons the material synthesized by this *in vitro* coupled ribosomal system includes the pea seed reserve globulin. Chromatin from regions which do not synthesize pea seed globulin *in vivo* does not support the synthesis of such globulin by the isolated ribosomal system. But the removal of histone from pea bud chromatin yields a DNA which will support globulin synthesis (BONNER, HUANG, and GILDEN 1963). This argues that the genes for globulin synthesis are repressed in the presence of histone. That histones are potent inhibitors of DNA-dependent synthesis of RNA *in vitro* establishes them as general inhibitors of the RNA-polymerase system. But this does not in itself show that histones function in a selective way in regulating the type or message content of the RNA produced *in vivo*. Native histone molecules, as they occur in the nucleohistone component of pea bud chromatin, contain RNA molecules chemically linked to them. This histone-bound RNA constitutes a new class of RNA, differing from other RNA types in base composition and chain length (Table 23). It is approximately 40 nu-

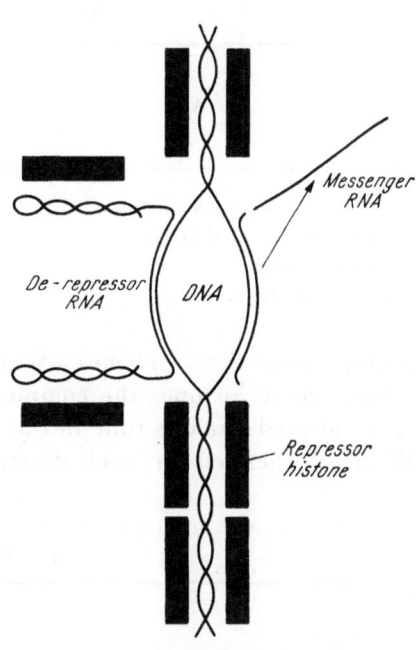

Fig. 32. Model to demonstrate the de-repression of genetic material as envisaged by FRENSTER (1965). Displacement of repression histones allows intermittent strand separation of the double helix. Specific de-repressor RNA then hybridizes with one of the DNA strands so freeing the complementary DNA strand for specific m-RNA synthesis.

cleotides in length and has a high content of dihydrouridylic acid (HUANG and BONNER 1965).

Now, addition or removal of histones also affects the amount of RNA synthesized by nucleoli. HUANG, BONNER, and MURRAY (1964), for example, have reconstituted soluble nucleohistones from individual histone fractions and a common DNA preparation by gradient analysis. The reconstituted complexes differ in their ability to support DNA-dependent RNA synthesis by both pea chromosomal and *E. coli* RNA polymerase. The histone of lowest arginine-lysine ratio yields the nucleohistone least active in support of RNA synthesis (Table 24). More significantly, the composition of the RNA produced can be altered in a consistent way. For example LIAU, HNILICA, and HURLBERT (1965) have used nucleoli isolated from the Novikoff ascites rat tumour to determine the amount and composition of newly synthesized RNA under *in vitro* conditions. The synthesis of RNA in these isolated nucleoli can be considerably influenced by the addition of histones.

For example, up to 90% of the RNA synthesis was inhibited by the addition of calf thymus histone. Moreover, the composition of the RNA that was formed was altered in the direction of a higher $(C + G)$ content (Table 25). The nucleolar proteins of Novikoff tumour cells themselves

Table 23. *A Comparison of the Nucleotide Composition of Different Species of Pea Bud RNA.* (After Huang and Bonner 1965.)

RNA Species	Mole Percent					
	C	A	U	DiHU	U + DiHU	G
1. RNA of whole nucleohistone	14.0	30.6	—	—	38.6	16.7
2. Histone-associated RNA	10.4	31.6	15.2	27.5	42.7	15.3
3. Transfer RNA	18.5	23.5	—	—	30.6	27.4
4. Ribosomal RNA	22.3	24.3	22.0	—	—	31.4

contain some 32% histones. When nucleoli were treated with trypsin to remove these histones the composition of the RNA produced by them was again altered but this time in the direction of a lower $(C + G)$ content. The RNA synthesized in nucleoli stripped of basic proteins approached the

Table 24. *Activities of Various Reconstituted Nucleohistones in Support of RNA Synthesis.* (After Huang et al. 1964.)

DNA Provided as	Histone/DNA Mass Ratio	Arg./Lys Ratio	RNA Synthesized (in μμMoles of Nucleotide per 10 min)	
			Pea Embryo Chromosome RNA Polymerase	E. Coli RNA Polymerase
Nucleohistone I b	1.37	0.1	0	56
II b	1.32	0.6	24	140
III	1.45	1.4	80	—
IV	1.35	1.4	216	4000
Whole calf thymus nucleohistone	1.33	0.49	0	—
DNA alone			320	8474

composition of DNA while that produced in the presence of added histones was more comparable to that of ribosomal RNA. This evidence suggests that nucleolar histones are involved in the regulation of the readout of DNA to produce r-RNA.

From the cases so far dealt with we see clear evidence for distinguishing between the classes of chromatin in interphase nuclei. In the prostate, four classes of interphase chromatin have been recognized (Liao and Lin 1967):

(i) Masked (M) regions. These occupy some 70–80% of the total nuclear DNA and they are not accessible to either RNA-polymerase or actinomycin D.

(ii) Restricted or repressed (R) regions. Occupying some 20–30% of the total DNA these regions are available for actinomycin binding and can function as a template in RNA synthesis by exogenous bacterial polymerase when the nucleus is ruptured. Prostatic RNA polymerase in the isolated nuclei does not, however, transcribe the DNA of this region *in vitro*.

(iii) Active regions. These form some 1% of the total nuclear DNA and in them DNA can be transcribed *in vitro* by the RNA polymerase associated with the prostatic nucleus. This RNA synthesis is not influenced

Table 25. *Effect of Removal or Addition of Histones on the Composition of RNA newly synthesized in Isolated Nuclei of Novikoff Ascites Rat Tumour Cells.*
(Data of LIAU et al. 1965.)

Base Ratio	RNA Synthesized *in vitro* Following				Normal Composition		
	Pre-incubation with Trypsin (μg/ml)			Addition of Whole Calf Thymus Histone	Nucleolar RNA*	Ribosomal RNA	Chromosomal DNA
	0	25	50				
C + G/A + U (T)	1.34	1.25	1.09	1.74	1.88	1.87	0.79

* The nucleolar proteins in Novikoff nuclei normally contain 32% histone.

by the level of androgenic steroid and is relatively insensitive to low concentrations of actinomycin D.

(iv) Nucleolar chromatin. This also occupies some 1% of the total DNA. RNA synthesis here is enhanced some seven-fold by the administration of testosterone to castrated rats and is extremely sensitive to a low concentration of actinomycin D both *in vitro* and *in vivo*.

One of the central problems of cell differentiation is how many different kinds of molecules single cells can synthesize concurrently. Many differentiating cells do not synthesize DNA at the time they are actively engaged in the process of differentiation. This appears to be the case in striated muscle, in skin and in many glandular tissues such as the pancreas and the mammary gland. For example the multinucleate myotubes which characterize skeletal muscle fibres originate by the fusion of mononucleated myoblasts. Synthesis of contractile proteins is confined primarily to the multinucleated myotubes. Thus those mononucleate cells which incorporate ^{3}H-TdR fail to bind antibodies against myosin and actin. On the other hand mononucleate cells or multinucleate myotubes which bind antibodies have G_1 nuclei and fail to incorporate ^{3}H-TdR. It would appear therefore that presumptive myoblasts repress pathways leading to DNA synthesis and withdraw from the mitotic cycle prior to transcribing for myosin (OKAZAKI and HOLTZER 1966). And myoblasts do not divide once they begin to synthesize contractile proteins. In a case like this where DNA synthesis appears to be turned off prior to commencement of the synthesis of cell-specific proteins, the cells concerned have a highly developed, submicroscopic, cyto-architecture which must be assembled prior to the commencement of rapid protein synthesis. DNA synthesis and subsequent cell division would of course disrupt this complex. The pattern of growth in such organs is therefore based on a "stem line" which undergoes rapid cell division and so produces a population of non-dividing cells which then undergo differentiation.

The growth of other tissues, however, does not involve stem lines. Rather both differentiating and differentiated cells are capable of division. This appears to be true of pigmented retina, cartilage and cardiac muscle. Here expression of function is not incompatible with rapid growth. Thus, chondrocytes derived from embryonic chick sterna, passaged several times in culture and grown as clones at a low plating density synthesize significant amounts of DNA and chondromucoprotein simultaneously (CAHN and LASHER 1967).

IV. Nuclear Inactivity

1. Total Inactivation

The formation of mature erythrocytes depends on a long developmental sequence of erythropoietic cell types which may be summarized as follows: Haemocytoblast or stem cell (HCB) → proerythroblast (PrE) → basophilic erythroblast (BE) → polychromatophilic erythroblast (PE) → normoblast (NORM) → reticulocyte (RET) → mature erythrocyte (ME).

GRASSO, WOODARD, and SWIFT (1963) followed the cytochemical changes in nucleic acids and proteins during erythrocytic development in the liver of the rabbit. The amount of cytoplasmic RNA per cell was initially high in the stem cell (HCB). But in successive stages of development there was a progressive decrease in the amount of RNA (Fig. 33). Using ^{3}H-CdR, a marked incorporation into RNA was observed in the HCB, PrE, and BE phases. All later stages showed no incorporation. This presumably reflects the lack of ribosomal RNA synthesis in erythroblasts which, in turn, results in a gradual loss of cytoplasmic RNA. Total cytoplasmic protein, in the cells used also for RNA determinations, decreased between the HCB and PE stages and then decreased sharply to the reticulocyte (RET) stage. This initial decline was, however, followed by an intensive synthesis of haemoglobin (Fig. 34). Nuclear protein measurements showed a decline throughout the entire erythropoietic process. This was accompanied by a decrease in nuclear area. The most active period of haemoglobin synthesis thus occurs in the later stages of erythropoiesis when the cytoplasmic RNA is relatively low.

Now late normoblasts have a 2 C-DNA level. A similar value is found in nuclei which appeared to be in the process of extrusion (Fig. 34). The DNA evidence thus indicates that the various erythroblast stages provide a means of proliferation. After the BE stage the nucleus plays, at the most, a minor role in the further differentiation of the system. Indeed, marrow cells, *in vitro,* exhibit a tendency to extrude their nuclei prior to active haemoglobin synthesis but incorporation of haem precursors continues to take place.

Differentiation of the erythrocyte thus depends on the establishment of a metabolically stable informational RNA complex in the earliest stages of the red cell lineage. After the establishment of this complex there is a loss of RNA synthesis and a gradual decrease in cytoplasmic RNA levels. At a point when the RNA content is relatively low an intensive haemo-

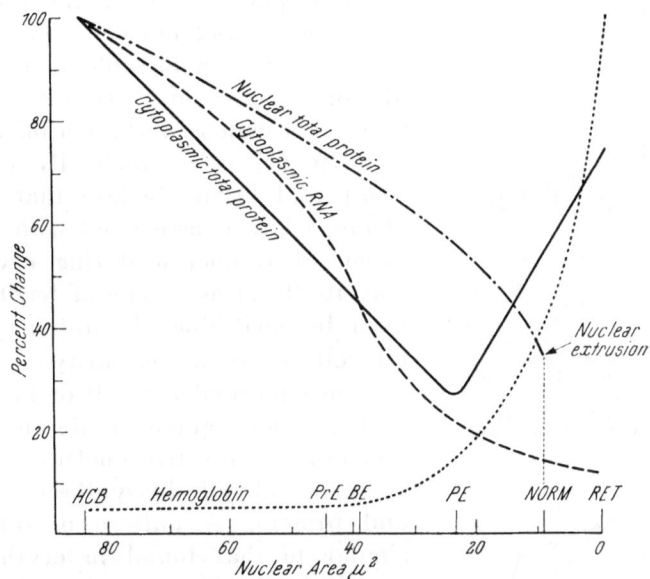

Fig. 33. Cytochemical changes involved in red blood cell production in the liver of the rabbit. HcB = Haemocyto-blast, PrE = Proerythroblast, BE = Basophilic erythroblast, PE = Polychromatophilic erythroblast, NORM = Normoblast, RET = Reticulocyte. (After GRASSO, WOODARD, and SWIFT 1963.)

globin synthesis is initiated. The nucleus would thus seem to serve a dual role. In the earliest stages of the process it produces a stable informational complex. In the later stages it provides a means of proliferation. Following this it becomes mitotically inactive and is expelled from the cell.

In birds, unlike mammals, the nucleus is not extruded during erythrocyte formation. Consequently all mature red cells are nucleated. In birds treated with ³H-UdR the average grain count per cell drops rapidly between the mid and late polychromatophilic stages and continues to decrease until no incorporation is found in the mature erythrocyte (CAMERON and PRESCOTT 1963). Protein synthesis as measured by ³H-leucine incorporation increases through the late polychromatophilic stages and is still high in the reticulocyte stage. In the mature erythrocyte, however, it falls to zero (Fig. 35). The nucleus of the mature erythrocyte is thus unique. It is not engaged in any synthesis associated with self-replication and its entire genome is permanently switched off.

In mature erythrocytes of the newt *Triturus cristatus* the chromosome regions contain a high concentration (about 45% of that of the cytoplasm) of a haem compound, probably haemoglobin, for, as in the case of birds, the nuclei actually contain haemoglobin. During erythropoiesis the nucleus of the developing erythrocyte undergoes a gradual condensation while the concentration of haemoglobin in the cytoplasm increases. Haemoglobin is also present in the nucleus during development and TOOZE and DAVIES (1963) have suggested that nuclear condensation is due in part to the interaction of haemoglobin and nucleohistone. Haemoglobin is known to be able to combine with phosphate ions and neutralize their negative charges. Condensed chromatin, as we have seen (p. 65) is metabolically inactive and TOOZE and DAVIES believe that the haemoglobin, which is associated with the condensation of chromatin during erythropoiesis, may itself act as a type of feedback mechanism for switching off protein synthesis as the cell approaches maturity.

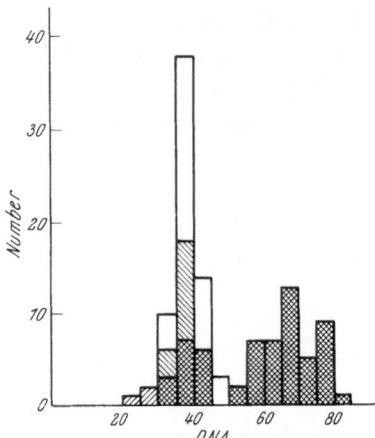

Fig. 34. Variation in DNA content of the nuclei of developing red cells of the rabbit. Solid bars = a mixed population of erythroblast nuclei exclusive of late normoblast stages, Open bars = late normoblast nuclei, stippled bars = nuclei in the process of extrusion, cross-hatched bars = extruded nuclei. (After GRASSO, WOODARD, and SWIFT 1963.)

The nucleated red cell of birds and amphibians thus cannot divide and they carry a completely inactive nucleus as manifested by its inability to biosynthesize DNA, RNA and protein. A pattern of differentiation similar to that found in erythropoiesis is found also in the conversion of the lens epithelial cell to an elongate fibre cell. This too is an example of terminal cell differentiation, in this case accompanied by the synthesis of γ-crystallin (STEWART and PAPACONSTANTINOU 1967). Here, as in the erythrocyte, maturation is associated with:

(i) loss of nuclear activity,

(ii) stabilization of m-RNA, and

(iii) a decrease in the polysomal population.

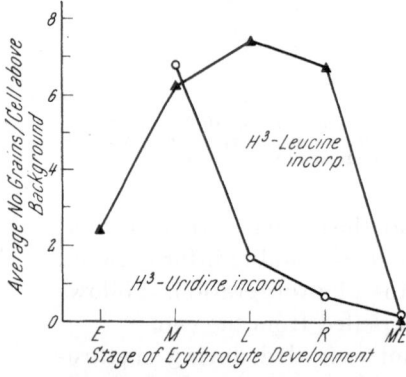

Fig. 35. Average grain counts at each stage of the erythrocyte series of the chicken 45 min after the injection of ^{3}H-UdR (—○—) and ^{3}H-leucine (—▲—). E = early polychromatic erythroblast, M = mid polychromatic erythroblast, L = late polychromatic erythroblast, R = reticulocyte, ME = mature erythrocyte. (After CAMERON and PRESCOTT 1963.)

2. Reposition

In the sperm nucleus the entire haploid set is reduced to a metabolically inert and minimal structure for this cell type is specialized for the transmission of the genotype between generations. During the conversion of the spermatid into the sperm (spermiogenesis), alterations occur in the nucleoprotein content of the nucleus. First, all RNA is lost from the nucleus and the DNA-associated histone is frequently replaced by a new type of protein characterized by a higher basicity and rich in arginine (histone shift or replacement). In the frog there is no such shift (BLOCH 1963). Neither does it occur in the one plant, *Tradescantia*, which has been examined from this point of view (RASCH and WOODARD 1959).

Histone replacement, where it occurs, is accompanied by an incorporation of ^{3}H-arginine suggesting that the change reflects synthesis of a new histone rather than the conversion of an existing protein (see also p. 76). In some species yet a further change occurs at the very end of spermatid conversion when the arginine-rich histone is in turn replaced in the definitive sperm by protamine. These protamines range in molecular weight from 4,000–10,800 and owe their strongly basic character to the presence of a

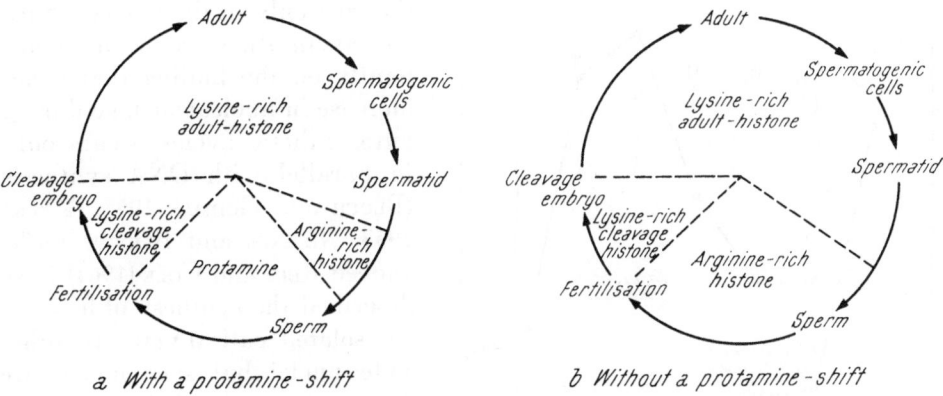

a With a protamine-shift   *b Without a protamine-shift*

Fig. 36. Changes in basic nuclear protein during the life cycle of (a) *Helix aspersa* (after BLOCH and HEW 1960) and (b) *Drosophila melanogaster*. (After DAS, KAUFMANN, and GAY 1964.)

very high proportion of arginine. Indeed more than 90% of their protein nitrogen may be present as arginine. Finally, unlike the histones which contain most of the common amino acids, protamines have relatively few and none of them are aromatic.

Such a protamine shift is found in *Loligo opalescens* (BLOCH 1962), *Helix aspersa* (BLOCH and HEW 1960) and in the bull (GLEDHILL et al. 1966). In the grasshopper *Chortophaga viridifasciata* (BLOCH and BRACK 1964) there is no protamine shift. Neither is there such a shift in *Drosophila melanogaster* (DAS, KAUFMANN, and GAY 1964) or in the mouse (MONESI 1965). This difference in synthetic behaviour between organisms with and without a protamine shift is summarized in Fig. 36.

In *Chortophaga viridifasciata* the process of cell elongation which occurs during spermiogenesis takes place in two stages. The initial phase involves the cytoplasm while the nucleus changes from a spherical to a highly elongate form in the second phase. The transformation of the nuclear proteins occurs during this second phase (Figs. 37 and 38). The post meiotic spermatid is very rich in RNA but its synthesis virtually ceases by the end of the first stage of elongation and all detectable RNA has been lost from the nucleus by this stage. Indeed RNA is also lost from the cytoplasm since most of this is, in fact, sloughed off at this stage (Fig. 38). Subsequent nuclear elongation is accompanied by a replacement of the typical lysine-rich histone by others rich in arginine (Fig. 37). Arginine incorporation occurs first in the thin cytoplasmic layer now surrounding the nucleus. Bloch and Brack (loc. cit.) have shown that this layer contains aggregations of ribosome-like particles. They therefore suggest that histone is synthesized in association with these RNA-units in the cytoplasm and then migrates into the nucleus where it combines with the DNA. Coincident with this, of course, the old histone is broken down, passed into the cytoplasm and sloughed off.

The suggestion that the histone shift depends upon cytoplasmic synthesis runs contrary to the assumption that histones are synthesized only in the nucleus. Such an assumption rests predominantly on the finding that a net increase in nuclear histone during autosynthetic cycles occurs only in parallel with DNA synthesis (Bloch and Godman 1955, Alfert 1958, Niehaus and Barnum 1965). Indeed Reid and Cole (1964) have described the synthesis of histones in isolated calf thymus lymphocyte nuclei but such nuclei are known to retain their own ribosome systems and so may be exceptional in this respect. Robbins and Borun (1967), by combining pulse-chase labelling of cells in G_1 and S with double labelling with tryptophan and lysine, have also shown that in HeLa cells histones are made in the cytoplasm on small polysomes. And Cave (1967) too has given evidence that histone synthesis can be uncoupled from DNA synthesis. High concentrations of non-histone protein and lysine-rich histones are localized in heterochromatic regions. Such regions are characterized by asynchronous DNA replication (see p. 22) but they do not show any marked differences with respect to ^{3}H-lysine incorporation. In special cases DNA-synthesis can precede complexing with histone by a substantial margin. Thus the doubling of DNA in the late replicating X-chromosome of the Katydid *Rehina spinosa* is virtually completed before histone doubling begins (Teng et al. quoted in Bloch et al. 1967).

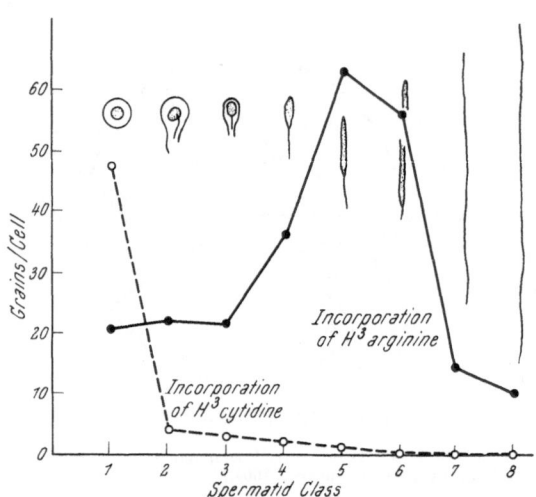

Fig. 37. Incorporation of H³-CdR into RNA and H³ argine into protein during the spermiogenesis of *Chortophaga viridifasciata*. (After Bloch and Brack 1964.)

The most complete quantitative study so far made on changes in deoxyribonucleoprotein (DNP) during spermiogenesis is undoubtedly that of Gledhill, Gledhill, Rigler, and Ringertz (1966). By the application of cytochemical methods on developing bull sperm they were able to show that:

(i) A drastic reduction in the Feulgen reactivity of nuclear DNA occurs

Table 26. *Microspectrophotometric Analyses of Nuclear DNA, DNP–PO₄⁻ and Basic Protein during Spermiogenesis of the Bull.*
(After GLEDHILL et al. 1966.)

Cell Type	DNA				DNP–PO₄⁻				Basic Protein			
	Total Extinction at 2650 Å after RNase + Cold TCA		Total Extinction at 5460 Å after FEULGEN		Fluorescence at 5300 Å after Acridine Orange		Total Extinction at 5900 Å after Methyl Green		Total Extinction at 5920 Å after Bromophenolblue		Total Extinction at 4800 Å after Sakaguchi	
	No. Cells	Mean	No. Cells	Mean	No. Cells	Mean	No. Cells	Mean	No. Cells	Mean	No. Cells	Mean
1. Round spermatids	17	7.10	200	7.72	43	3.70	10	4.44	24	0.46	19	1.53
2. Elongate spermatids	18	6.71	200	7.64	10	1.49	10	2.64	26	1.56	20	1.45
3. Testicular spermatozoa	30	6.50	119	4.20	10	0.21	2	1.43	22	6.18	20	3.27
4. Caput epididymal spermatozoa	150	6.68	360	3.52	40	0.10	—	—*	20	6.97	20	3.54
5. Ejaculated spermatozoa	202	6.61	355	2.63	40	0.10	—	—*	20	7.02	10	3.63

* Too small to measure.

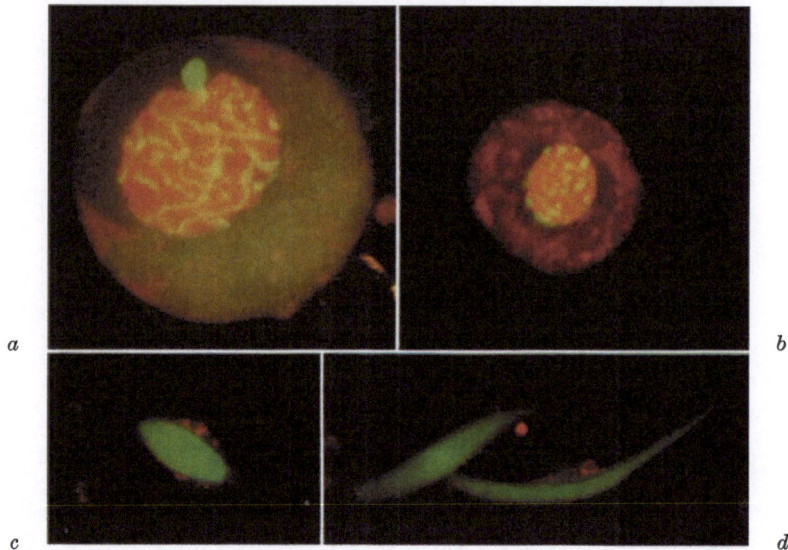

Fig. 38. Fluorescence patterns following treatment with acridine orange (1 : 10,000) in *Chortophaga viridifasciata*. (*a*) is a cell while (*b*), (*c*) and (*d*) are stage 1, 4 and 5 spermatids respectively. Notice that the X-chromosome in (*a*) and the entire nucleus in (*d*) are bright green indicating the absence of RNA. (After BLOCH and BRACK 1964.)

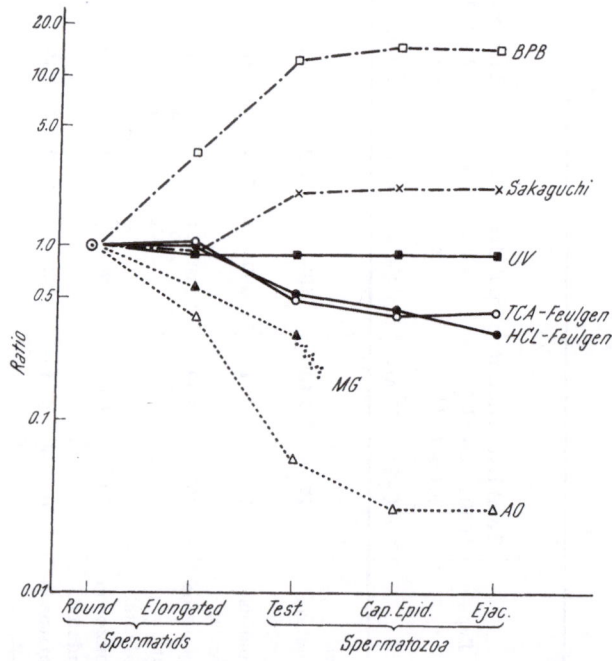

Fig. 39. Summary of cytochemical changes during sperm formation in the bull (after GLEDHILL et al. 1967). For each cytochemical parameter the mean values for the different cell types have been related to the corresponding mean values for "round" spermatids to yield a ratio which has then been plotted on a log scale (ordinate).

when the elongate spermatids differentiate into testicular spermatozoa. This reduced reactivity is not due to a reduction in the amount of DNA itself since the nuclear absorption at 2,650 Å is unaltered throughout spermiogenesis (Table 26).

(ii) A marked increase occurs in the basicity of the nuclear protein as measured by the alkaline bromophenol blue (BPB) technique. The results obtained with the Sakaguchi reaction likewise indicate an increase in the nuclear content of protein-bound arginine.

(iii) Concomitant with the decrease in Feulgen reactivity and the increase in protein basicity there is a decrease in the total number DNP-PO$_4$ groups available for binding by basic dyes. This was reflected by the decreased ability of DNP to bind acridine orange or methyl green (Fig. 39) and indicates an increase in the strength of electrostatic binding of the nuclear protein to the DNA molecule.

V. Mechanical Nuclear Activity

1. The Coiling Cycle

The transformation of the later interphase chromosome to the metaphase state depends on a process of spiralization and condensation. When fully despiralized an interphase chromosome might be expected to be at least as long as the lampbrush chromosomes of the newt (see p. 53), which means in excess of 1 mm. By contrast chromosomes at metaphase are usually 10 μ or less in length. Thus length changes of the order of a hundred fold must be involved.

ANDERSON (1956) has developed a model of coiling based on the following considerations:

(i) In the chromosome, chains of DNA are linked together at intervals by basic protein in such a way that fairly long interhistone lengths of DNA exist.

(ii) Non-histone proteins are loosely bound on both the histone and the DNA.

(iii) In the "extended" state most of the DNA phosphate groups are not close to the positively charged groups of the histones. If the loosely held protein is removed, the interhistone DNA segments will, at the ionic strengths believed to exist in the cell, tend to coil and cross-link with adjacent histone molecules. This results in condensation and contraction. Similar results may obtain from the addition of polycations. Indeed ANDERSON believes that polyvalent cations are responsible for chromosome condensation.

The essence of ANDERSON's argument is that chromosomal condensation is due to configurational changes in DNA molecules dependent on the discharge of negative groups on DNA by polyvalent cations. ANDERSON's theory implies that the volume occupied by chromatin material is controlled by competition between acidic and basic proteins, or other polycations, for sites on the nucleohistone. When the balance between these competing substances is shifted in favour of the polycations then the DNA chains coil because more of the surface of the DNA and the histone molecules become available for mutual association.

Support for this type of argument comes from a number of different directions:

(i) Experimental condensation of chromosome-like bodies can be obtained in isolated rat liver nuclei by the addition of the dibasic amino acid arginine (WILBUR and ANDERSON 1951).

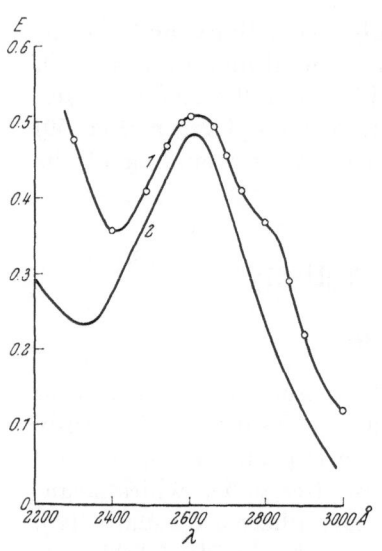

Fig. 40. Absorption spectrum of the central part of a mitotic metaphase chromosome of *Chorthippus* (curve 1) as compared with that of pure nucleotide (curve 2). Note the presence of considerable amounts of protein as indicated by high absorption at 2,300 Å together with tyrosine-tryptophan bands at 2,800 Å. (After CASPERSSON 1956.)

(ii) The neuroblast cells of *Chortophaga viridifasciata* are large (25 μ in diameter), rapidly dividing (total cycle time of 6 hours at 38° C), cells which can be grown and studied *in vitro*. Three mitogenic agents (kinetin, agmatine and hypertonic culture medium) are known which induce mitosis in cells of normally non-dividing tissue and/or accelerate interphase and mitosis in cells of dividing tissue. An 0.005 M solution of agmatine (decarboxylated arginine) accelerates the rate at which cultured neuroblasts go through pro-metaphase and metaphase by some 32% (ST. AMAND et al. 1955). Significantly agmatine is a polycation with two negative charges.

Again, when neuroblasts are cultured in a medium containing 1.2 times the regular inorganic salt concentration, accelerated contractions occur within 30 seconds of the cells being placed in the hypertonic medium. There is evidence which implicates divalent ions, especially those of Ca^{++} and Mg^{++}, in chromosome organization (STEFFENSEN 1953, 1955). The accelerated contraction produced by hypertonic medium might therefore be attributable to the increased amounts of such ions in the medium. Significantly the binding between DNA and histone would be strengthened by divalent ions.

(iii) CASPERSSON (1956) has shown that one of the most marked changes which occurs during prophase is a decrease in the amount of protein in the nucleus. Moreover, the protein that remains is arginine-rich (SERRA 1947) and, in fact, represents most of the basic protein of the cell. Thus, absorption curves of metaphase chromosomes reveal absorption bands at 2,300 Å in addition to tyrosine-tryptophan bands at 2,800 Å (Fig. 40). Clearly, metaphase chromosomes contain considerable amounts of protein despite the loss of nuclear protein. Indeed, MAIO and SCHILDKRAUT (1966) have shown that the DNA/protein ratio in isolated metaphase chromosomes is 1 : 5.

(iv) Isolated DNA-histone particles double their length when the histone is removed (MARMUR and DOTY 1961). And, as we have seen earlier (p. 65), removal of lysine-rich histones from isolated calf thymus nuclei causes dense chromatin to become diffuse. Moreover, when protein synthesis is blocked by selective inhibitors the chromosomes are less contracted (Table 27).

Table 27. *The Effects of Various Treatments on Chromosome Condensation.*

Nuclear Type	Treatment	Effect	Reference
Interphase	Arginine	Induction of prophase-like stage	ANDERSON 1956
	Ionic elevation, basic proteins, reduced pH		PHILPOT and STANIER 1957
	Colcemid	Late replicating segments incapable of coiling normally	STUBBLEFIELD 1964
Mitotic	Polyamines	Suppression of anaphase coiling	DAVIDSON and ANDERSON 1960
	Phenethyl alcohol		BAMMI and JURA 1966
	Versene, sodium citrate, urea	Relaxation of, or in the case of urea complete, uncoiling	TROSKO and BREWERN 1966
First Meiotic prophase	NaCl, KCl, CaCl$_2$	Abnormal spiralization	MATSUURA and IWABUCHI 1962
	Ammonia, KCN	Uncoiling	KUWADA et al. 1938
	Base or amino-acid analogues 8-azaguanine 5-fluorouracil ethionine and p-fluorophenylalanine	Uncoiling of diakinesis chromosomes	KEMP 1964
	Actinomycin D mitomycin C 5-fluorouracil ethionine p-fluorophenylanine and 5-methyltryptophan	Uncoiling of anaphase I chromosomes	

(v) Sulphur, which is absent from DNA, is a constituent of most proteins. Selenium will substitute extensively for sulphur in protein. The selenohydryl group is less reactive than the sulphydryl and this, together with associated differences in bond strength and distance, may be expected to lead to alterations in the physicochemical properties of seleno-substituted proteins. Cytological examination of meiosis in barley plants treated with sodium selenate at the seedling stage showed a reduction in chromosome coiling (Walker and Ting 1967).

To what extent other materials play a role in chromosome coiling is not known. As we have seen (p. 44), RNA is taken up by the chromosomes in late prophase. And, at about the time that spiralization begins, phospholipids are deposited on the chromosomes (La Cour and Chayen 1958). Stubblefield (1964) found that late replicating segments of hamster chromosomes entering G_2 apparently prior to the apparent completion of DNA synthesis fail to contract in normal fashion and reach the metaphase stage in an extended state. If the proteins of the chromosome do aid in contraction this would suggest that completed replication is a necessary condition for successful protein attachment.

2. The Movement Sequence

During the autosynthetic cycle, whether this is mitotic or meiotic in character, the mechanical phase of chromosome movement is associated with and dependent upon the development of a transitory and highly orientated spindle system. Under normal conditions the spindle cycle is precisely correlated with the chromosome cycle but the two sequences can be divorced experimentally. For example, colchicine, at the proper concentration, can disorganize the spindle. In such circumstances the chromosome cycle is not disturbed except for a supercontraction process.

The extent to which the development and activity of the spindle system depends on the action of chromosomes themselves remains to be fully clarified. Clearly the bulk of the genetic material is incapable of biosynthetic activity during the division sequence itself (see p. 46). The centromere, however, is uncoiled at this time and significantly it is the centromere that is concerned with chromosome-spindle relationships. Centromeres are displayed with great clarity in E.M. sections of colcemid-treated chinese hamster fibroblasts following double fixation in glutaraldehyde and OsO_4 and subsequent embedding in Epon (Brinkley and Stubblefield 1966). With this technique the centromere appears to consist of a dense core surrounded by a less dense zone in which numerous microfibrils loop out at right angles to the axial fibrils of the core in a lampbrush-like fashion.

Micromanipulation and centrifugation studies confirm the view that the spindle is a semi-solid composed of very wet and weakly associated materials of higher density than the surrounding cytoplasm and organized into highly elastic, oriented fibres. The electron microscope reveals that these fibres are, in fact, microtubules which range from 130—270 Å in diameter.

The principal chemical component of the spindle system is protein. For example in the unfertilized egg of the sea-urchin *Strongylocentrotus* there is a total of 6.2×10^{-5} mg of protein. That in the mitotic apparatus of the same organism amounts to 0.72×10^{-5} mg. This is about 11.6% of the total cell protein and is some fifty times as much as that of the nucleus. The spindle protein is of the ribonucleoprotein type and its average molecular weight is 315,000 ± 20,000 (Zimmerman 1960).

A large amount of material is thus involved in the construction of the spindle and there is evidence that this material is present some time before the actual spindle is organized. WENT (1959) undertook an immuno-chemical characterization of spindle protein with the principal aim of deciding whether the spindle system was:

(i) synthesized *de novo* from very small, immunologically non-specific amino-acids or small polypeptide units, or

(ii) formed as a result of the assembling, with little or no structural modification, of pre-existing molecular units.

To this end he prepared rabbit anti-sera against the spindle and the unfertilized egg proteins of the eggs of sea-urchins. The spindle was found to contain two main antigens which were designated precursor-1 and pre-cursor-2. Both of these, together with others which were not present in the spindle, were found in the unfertilized egg which, in sea-urchins, com-pletes meiosis prior to sperm entry. Consequently the second alternative (ii above) was preferred.

This investigation shows that, in sea-urchin eggs at least, the spindle material in more-or-less its final form is produced some time before the spindle is actually organized. A pre-metaphase build-up and partial organization of spindle material can be demonstrated also in endosperm cells. Here a distinct clear region develops around the nucleus at prophase. By late prophase this region is spindle-shaped which presages the orienta-tion of the spindle itself. Indeed the material it contains is oriented and so the clear zone exhibits birefringence prior to the breakdown of the nuclear membrane (BAJER 1957).

The primary spindle structure that forms at the disruption of the nuclear membrane consists of a system of continuous fibres which run from pole to pole. Once the chromosomes meet this primary structure their centromeres are responsible for the secondary organization of the system. More specifically the centromeres accumulate microtubules into dense bundles of chromosome fibres (half spindle fibres) which run from each sister half centromere (mitosis and meiosis-II) or from paired, homologous centromeres (meiosis-I) to a spindle pole. That the centromeres themselves play a role in this process is supported by the fact that they can also organize fibres from the phragmoplast (MOLÈ-BAJER 1965). According to BAJER (1966) the microtubules attached to the half centromere pairs are supported and kept in a state of tension by the activity of the inter-tubular material.

Having established a relationship with the primary spindle the chromo-somes evince a strong tendency to move toward the equatorial region (con-gression) while all other bodies not attached to microtubules are moved toward the spindle poles (ÖSTERGREN, MOLÈ-BAJER, and BAJER 1960). This pro-metaphase movement is characterized by complete individuality of response on the part of each chromosome. By contrast the anaphase move-ment, which determines the final course of the division sequence, involves interactions both within and between chromosomes. The most complete study of these interactions is that of FORER (1966). By UV-microbeam irradiation of localized spindle areas in living spermatocytes of *Nephrotoma suturalis* $(2n = 6 + XY)$ during or shortly before anaphase-I Forer found:

(i) Chromosome movement is affected more strongly by half spindle irradiation than by the irradiation of the interzonal fibres which are evident between the two separating chromosome groups.

(ii) Following metaphase irradiation: (a) Both sister half-bivalents responded in identical fashion even though only one half spindle area in fact received irradiation. Thus, both half-bivalents temporarily stopped moving after irradiation and both resumed movement at the same time and subsequently moved with the same velocity. (b) In 18 out of 19 cases analyzed all three pairs of autosomal half-bivalents responded in an identical way to localized irradiation of part of one half spindle system. Moreover, the different half bivalents in the cell all resumed movement at the same time.

(iii) Following anaphase irradiation: (a) When one half-bivalent stopped moving in response to the irradiation of its half-spindle unit its sister half-bivalent stopped moving also. And when one half-bivalent slowed down its movement its sister responded in an identical fashion. (b) Unlike the response to metaphase irradiation, different bivalents in a given cell are independent in their response to irradiation at anaphase.

The interdependence of chromosome movements implicit in Forer's work is mediated, at least in part, by the functioning centromeres and imply an as yet unspecified aspect of chromosome function. That a chromosome can contribute to its own movement is demonstrated also by the behaviour of dicentric chromatid bridges produced by crossing over within X-chromosome inversion heterozygotes in female *Drosophila melanogaster*. Dicentric bridges derived from X-chromosome inversions normally stall on the spindle at both first and second divisions of meiosis. However, X-attached centromeres which are associated with a complete long arm of the Y-chromosome, either as an entire extra arm or else as an interstitial addition, are capable of breaking dicentrics both at first (Lindsley and Novitski 1958) and second (Ptashne 1960) division. Measurable differences in the kinetic activity of centromeres result from the constitution of the heterochromatic regions immediately adjacent to the centromere. The recent study of Camenzind and Nicklas (1968) on the pattern of non-random chromosome segregation at male meiosis in *Gryllotalpa hexadactyla* is also relevant to Forer's findings. Here the univalent X-chromosome consistently moves to the same pole at first anaphase as the larger member of a heteromorphic autosomal bivalent. This behaviour appears to be regulated via the X-chromosome which, though it shows no mechanical connection with the bivalent in question, is able to reorient if an abnormal segregation occurs spontaneously or is induced by micromanipulation.

VI. Synthetic Sequences and the Problem of Regulation

The pollen mother cells of *Lilium longiflorum* are a specialized group of cells which are antecedants of the pollen grains. During the life of the microspore, which lasts for several weeks, thymidine kinase activity appears at a precisely defined time and lasts no more than twenty four hours. (Fig. 41)

By using agents which distort RNA synthesis (8-Azaguanine, 5-Fluorouracil), interrupt the presumed translation of RNA information (chloramphenicol) or block or distort the pattern of aminoacid incorporation into protein (5-Methyl tryptophane and ethionine) it was shown that the appearance of TdR-kinase is dependent on the synthesis of RNA and

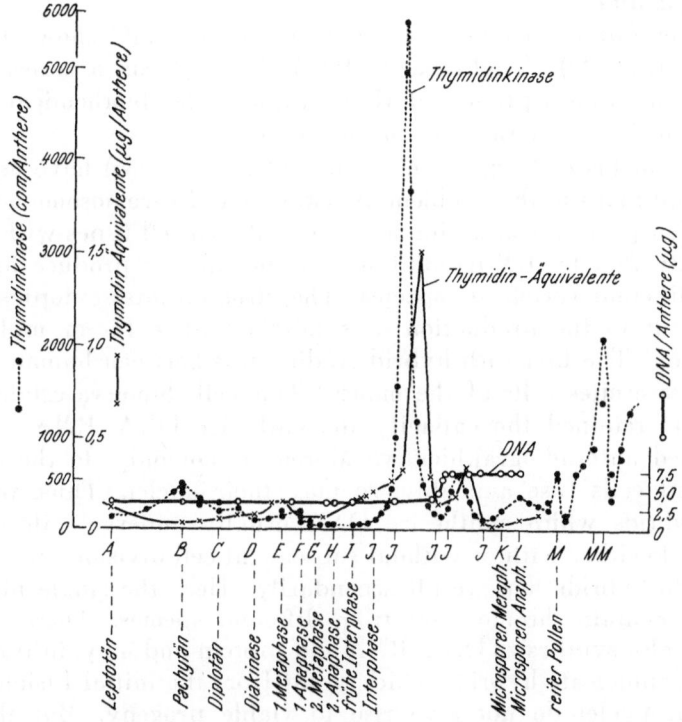

Fig. 41. The pattern of the post-meiotic synthesis of thymidine kinase in anthers of *Trillium erectum*. Also shown are the DNA-content of microsporocytes and microspores and the production of deoxyribosides. (Data of HOTTA and STERN 1961 as presented by DUSPIVA 1966.)

protein. All these reagents, if added prior to the normal appearance of enzyme activity, virtually abolish such activity (HOTTA and STERN 1963).

The brief appearance of TdR-kinase activity prior to DNA synthesis in the microspores of lilies (Fig. 41) thus appears to be due to a *de novo* synthesis of protein, the chain of events leading to such synthesis beginning with the production of RNA. HOTTA and STERN suggest that this sequence has all the earmarks of induced enzyme synthesis. They also suggest that mitosis and meiosis are composites of individually regulated molecular systems held together by a system of regulatory devices similar to that governing TdR-kinase synthesis. Certainly, there is a marked periodicity in many of the biochemical activities which characterize the microspores. Thus, transient peaks in the concentration of presumed nucleic acid precursors appear to precede the DNA syntheses which occur during the pre-

meiotic interphase of the PMC's, late microspore interphase and again at
the S-phase preceding the formation of the male gametic nuclei. Near the
time of each DNA synthesis a pool of deoxyribosides appears. In all probability the last pool represents the breakdown products of the DNA of the
tapetum. During the time of breakdown of tapetal tissue, enzymes which
degrade DNA into nucleotides and dephosphorylate these into deoxyribosides have been demonstrated in anther homogenates. Isolated microspores of *Trillium* also show a transient high activity in the phosphoryla-
tion of thymidine.

Again in *Lilium henryi* every enzyme of the PMC shows high phase
specificity (Fig. 42). To Linskens (1966) this suggests a series of phase-
specific gene transcriptions which coordinate the biochemical sequences
governing meiosis and the post-mitotic mitosis.

Harris and his colleagues (see review of Harris 1966) have used a quite
different approach to the problem of regulation of chromosome activity. By
treating a suspension containing a mixture of two cell types with an animal
virus inactivated by UV-irradiation it is possible to produce hybrid cells
between different vertebrate species. The virus initiates cytoplasmic fusion
and so leads to the production of syncytia with a bi- or multi-nucleate
organization. The first such hybrid studied was between human HeLa cells
and Ehrlich ascites cells of the mouse. This cell chimaera established that
both nuclei retained the capacity to synthesize DNA, RNA and protein
when tested autoradiographically. Moreover, not only do the cytoplasms
of different cells fuse amicably, so may their nuclei. Thus, those nuclei
in cell hybrids which synthesize DNA usually undergo mitosis and this
they may do either with or without subsequent cell division. In either event
uninucleate hybrids may result secondarily. Here the single nucleus will,
of course, contain chromosomes from different species. These uninucleate
chimaeras also synthesize DNA, RNA and protein and may, in turn, undergo
mitosis. Uninucleate hybrids which result from the initial fusion of a large
number of nuclei do not give rise to viable progeny. But those which
contain only a single diploid set of chromosomes from each parent are, in
some cases at least, capable of prolonged multiplication.

In clones of interspecific cell hybrids it is clear that the RNA transmitted from the hybrid nucleus to the hybrid cytoplasm does not lead to
extensive misreading of genetic instructions. Likewise, each set of chromosomes is perfectly able to respond to cytoplasmic signals of mixed origin.

During the course of differentiation certain vertebrate cells lose the
ability to synthesize DNA or RNA or both of these (see p. 72). For example,
small rat lymphocytes synthesize variable amounts of RNA, do not normally synthesize DNA but can be induced to do so by suitable antigenic
stimuli. Rabbit macrophages, on the other hand, synthesize RNA but not
DNA while the nucleated erythrocytes of the hen are incapable of engaging
in any forms of synthesis. By mixing these varying cell types with one
another and with human HeLa cells it is possible to test the synthetic
capacity of the different nuclei when placed against a chimaeric cytoplasmic background. In all cases the results are the same (Table 28):

(i) If either of the parent cells normally synthesize either DNA or RNA or both then the equivalent syntheses take place in both types of nuclei in the hybrid,

(ii) if neither of the parental cells normally synthesizes DNA then no synthesis of DNA occurs in the hybrid, and

(iii) the induction of synthesis is a response to signals emerging from the foreign cytoplasm. This is especially clear in the case of HeLa-erythrocyte hybrids. Here the virus has to be used at such high concentrations that it causes the erythrocytes to lyse prior to hybridization. The naked erythrocyte nuclei are then incorporated into the HeLa cytoplasm. In such hybrids the only signals which can stimulate the synthesis of DNA and RNA in the erythrocyte nuclei must, of course, come from the HeLa cytoplasm.

The regulation of nuclei acid synthesis in cell hybrids is thus essentially unilateral. Whenever a cell which synthesizes a particular nucleic acid is fused with one which does not, the active cell type initiates an equivalent synthesis in the inactive partner. But in

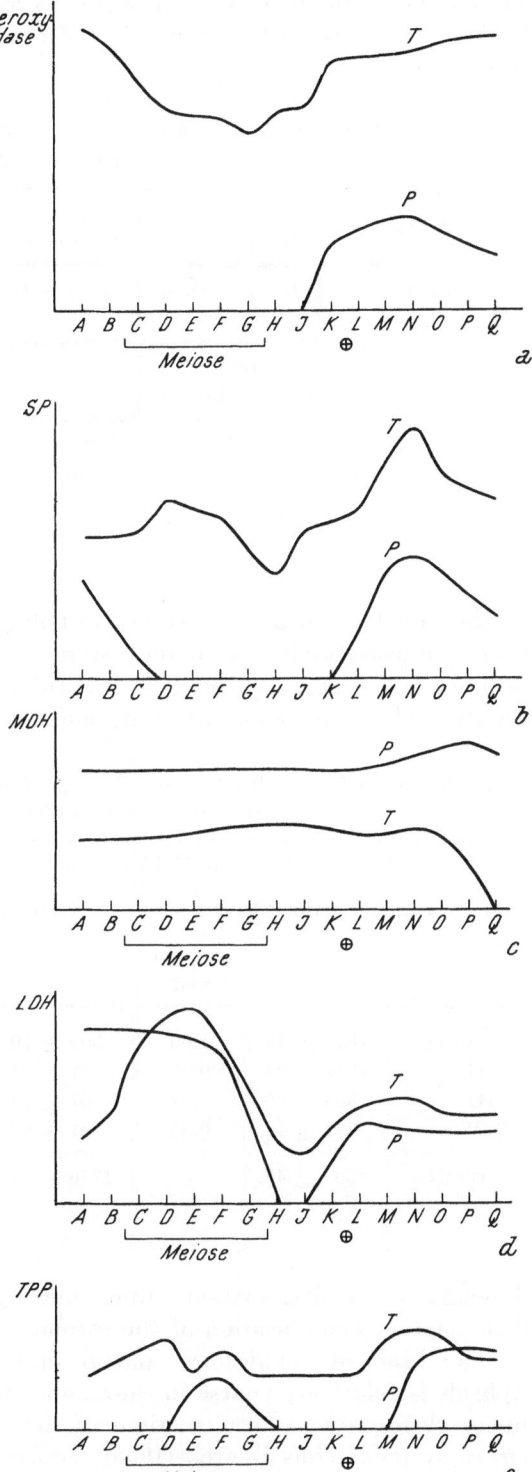

Fig. 42. Variations, shown in terms of arbitrary units, in the activity of peroxidase (a), acid phosphatase (b), malate dehydrogenase (c), lactic dehydrogenase (d) and thiamine pyrophosphate (e) during meiosis and pollen development in *Lilium henryi*. Note (+) marks the tetrad stage, P = pollen, T = tapetum. (After LINSKENS 1966.)

no case does the inactive cell suppress synthesis in the active one. This holds even in cases where a number of inactive cells are fused with a single active one.

Table 28. *Nucleic Acid Synthesis in Hybrid Cell Combinations.*
(After Harris 1966.)

Cell Type	Human HeLa	Rabbit Macrophage	Rat Lymphocyte	Hen Erythrocyte
Human HeLa	RNA + DNA	RNA + DNA	RNA + DNA	RNA + DNA
Rabbit Macrophage		RNA	RNA	RNA
Rat Lymphocyte			RNA	Not tested (Predict RNA)
Hen Erythrocyte				No synthesis

Now nuclei which are severely restricted in their capacity to synthesize RNA are usually small and they stain deeply with chromatin dyes. When such nuclei resume synthesis in hybrid cells they enlarge and stain less deeply. The conversion of a dormant nucleus into an active one thus

Table 29. *Distribution of DNA in the Salivary Gland Chromosomes of Chironomus tentans.*
(Data of Daneholt and Edström 1967.)

Chromosome No.	Amount of DNA per Chromosome				Relative	
	8192 C-Nuclei		4096 C-Nuclei		Length Relations	No. Chromosome Bands
	μμg	Relative Amt.	μμg	Relative Amt.		
I (X)	1044 ± 17	1.00	555 ± 10	1.00	1.0	1.0
II	1027 ± 24	0.98	532 ± 21	0.96	1.0	0.8
III	908 ± 20	0.87	487 ± 10	0.88	0.8	0.76
IV	361 ± 14	0.35	191 ± 8	0.34	0.3	0.34
Totals	3660 ± 49		1776 ± 41		Total length of haploid set = 700 μ	Total No. of bands = 1,900

depends in the first instance upon messages which regulate the degree of dispersion or condensation of the chromosome material (see also p. 81).

The kind of regulatory control implicit in the experiments on cell hybrids is relatively coarse in character. Such control is capable of opening up or closing down large sections of the chromosome system or indeed the entire system. Thus Burton (1968) reports that phytohaemaglutinin (PHA) induces mitogenesis in short term cultures of peripheral lymphocytes of the kangaroo through the removal of histone from the DNP complex. There is, however, evidence for systems of fine control too. As we have seen earlier some, at least, of the fluctuations and alterations which occur in the organization patterns of chromosomes reflect the role which the chromosomes play in the control of biosynthetic processes. In this connection, KEYL (1965), PELLING (1966), and BEERMANN (1967) have argued that the transverse band of dipteran polytene chromosomes and its counterpart, the chromomere, in normal chromosomes is a unit of replication and transcription.

DANEHOLT and EDSTRÖM (1967) working with the polytene system of *Chironomus tentans* have recently shown that (Table 29):

(i) the amount of DNA in a chromosome is a function of its length,

(ii) the average amount of DNA per band is similar for different chromosomes,

(iii) the total amount of DNA per chromatid set is 0.2 $\mu\mu$g which corresponds to a double helix length of some 500 mm, and

(iv) if DNA is considered to be shared mainly by the visible bands then each chromatid would contain something like 10^{-16} gm of DNA per chromomere.

If these calculations are accepted then a chromomere corresponds to something like 100,000 base pairs and could thus serve to code for about 30,000 amino acids. This is improbably high for the product of a single cistron and it suggests that either bands represent operational units of greater complexity than has hitherto been assumed or else a large part of the DNA must be non-functional in an informational sense. That is, there must be a considerable repetition of a given sequence. CALLAN (1967) has arrived at an essentially similar conclusion to the latter with regard to lampbrush chromosomes. He suggests that each unit of information encoded as a DNA base sequence is, in fact, serially repeated. In such sequences a terminal unit serves as a master copy which is then followed by a series of slave units. The master genes are believed to be distinct from the remainder of the genetic material in the sense that, in the germ line, RNA synthesis is the sole function of the slaves.

HESS (1967 a) has found several mutations affecting Y-chromosome loop morphology in irradiated males of *D. hydei*. Cytogenetic tests show that these mutations have occurred either within or very near to the affected loops. There is no indication of any gross structural change in the Y and the males carrying these mutations are fully fertile. To account for the observed change in morphology of the loop as a whole by an apparent point mutation BEERMANN (1967) has suggested that there is a functional subdivision within the loop locus such that an initial "master" segment which is active in transcription, is followed by a long segment with auxiliary functions. Given such a subdivision into transcribing and stabilizing or "packaging" segments it follows that a mutation in the master transcribing segment would be expressed as a change in the loop as a whole.

The relationship envisaged here would, of course, be highly specific in the sense that a given packaging segment would deal only with copies from its own initial transcribing segment. This is most easily accomodated if one assumes that the packaging segment represents a multiple redundant portion of the master segment. KEYL (1965) has described what he believes to be intra-chromomeric duplications in *Chironomus thummi thummi* (see p. 36) and such duplications would conceivably serve as a basis for producing and adding to packaging segments.

This idea of master and slave segments goes some way towards accounting for the occurrence of a large excess of DNA over and above that required to specify the structure of cell proteins. It also helps to reconcile the wide disparity in DNA value which exists between different species. Thus it is not self evident why a lily requires thirty times as much DNA as a lupin (MARTIN 1966) or why a salamander should have 20–30 times as much DNA as a man. Such wide variability in the quantity of DNA present in different species itself suggests that much of the DNA cannot contain exclusive genetic specifications for the synthesis of distinct proteins.

There are, however, complications to the arguments proposed above. In the case of the polytene chromosome for example, the apparent one-to-one relationship between chromomeres (bands) and genes is based on a comparison of the banding patterns in larval tissues with the genetic characters of the adult. And cytogenetic data do not tell us whether the units defined by genetic experiment actually involve the entire chromomere or only specific portions of it. As PELLING (1966) points out a chromomere in *Chironomus* is equivalent in molecular weight to the size of an entire phage "chromosome".

The function of the lampbrush chromosome of amphibians is even more obscure. Its appearance is especially deceptive for, at first sight, many if not most loci appear to be synthetically active. Despite earlier claims to the contrary the lampbrush chromosome appears to play little if any role in directing yolk synthesis since this occurs mainly in the liver (DAVIDSON and MIRSKY 1965). It certainly is involved in RNA synthesis but as we have seen (see p. 55) only 3% of the total genome appears to be involved. Of course the formation of RNA may be only part of the pre-programming of the egg cytoplasm for early development. In fact DAVIDSON and MIRSKY (1965) have given evidence to suggest that newly matured egg cytoplasm contains substances responsible for the arousal of specific genes in nuclei exposed to them, in other words that selective gene activators are present in the cytoplasm. Even so it is very difficult to explain why such a large number of distinct sites in the same chromosome and so many different chromosomes are involved in lampbrush activity. Indeed this must argue strongly against any simple regulatory system of the kind implicit in the master slave argument. Neither is it clear whether an inter-chromomeric segment of the lampbrush system is ever represented as a chromomere in any other cell type. Should this prove not to be so then there can be little doubt that, like the bands of polytene systems, many lampbrush chromomeres must represent compound genetic units rather than independent units of genetic action. And again we are faced with the problem of deciding what role, if any, the inter-chromomeric sections of the chromosome play. The minimum mean length of continuous DNA molecules isolated from salivary gland chromosomes of *Chironomus* is of the order of 50 μ and these are linear, not circular (WOLSTENHOLME, DAWID, and RISTOW 1969). EDSTRÖM (1964) has argued that the mean haploid DNA content of a polytene band in the largest salivary gland nuclei of *Ch. tentans* is $6 \times 10^{-5}\ \mu\mu$g, which is equivalent to a piece of DNA some 18 μ long. If this estimate is correct then a single molecule of DNA must, on average, involve at least 2—3 bands of the polytene unit. The mean length of DNA molecules from salivary gland chromosomes of *D. melanogaster* is 72 μ and here the mean haploid DNA content per band is equivalent to about 9 μ of DNA length (RUDKIN 1965). In this case a single DNA molecule would involve at least 7—8 bands and, of course, the intervening interband regions (WOLSTENHOLME et al. loc. cit.).

VII. Chromosome Structure

Although much is known about the roles and behaviour of the chemical components of chromosomes little is understood about the way these components are organized in the chromosome. The relationship between DNA molecules and "chromosome" organization is best understood in bacteria (protokaryotes). Here there is only one double DNA helix per genome. This has a total length of 700–900 μ and a molecular weight of 2.3–4.6 $\times 10^9$ daltons according to the state of replication (DAVERN 1966). The naked DNA is arranged in the form of a circle with a single site for the initiation of DNA synthesis which proceeds around the circle once for every cycle of bacterial replication (CAIRNS 1963). It takes about 30 min to complete replication and some 20–30 μ of DNA is duplicated per min.

On treatment with 0.5 N NaOH, dimeric DNA dissociates into single strands. Such treatment, in combination with a sucrose-centrifugation for a time sufficient to yield a DNA of constant molecular weight, allows the recovery of experimentally-produced single-stranded DNA. In T_2 phage such strands represent the entire "chromosome". McGRATH and WILLIAMS (1967) report, however, that DNA from *E. coli* dissociates into multiple, 10–12, single stranded segments when treated in this way. The significance of this dissociation relative to the organisation of a *coli* "chromosome" remains to be clarified.

The condition in plants and animals (eukaryotes) differs from this simple bacterial system in four fundamental respects:

(i) There are at least two and usually many more chromosomes per genome. These chromosomes are large by comparison with the bacterial units and contain many times the amount of DNA found in bacteria.

(ii) Histone molecules are regularly associated with the DNA.

(iii) The chromosomes go through distinctive division cycles (mitosis and meiosis) and are enclosed in a nuclear membrane between and indeed sometimes during such cycles, and

(iv) although there have been claims that the circularity of DNA is not confined to bacteria (HAYASHI et al. 1964, HOTTA and BASSEL 1965) the balance of evidence is that the nuclear DNA in plants and animals is linear in organization. On the other hand, highly twisted circles were found in a fresh preparation of mitochondria from oocytes of *Xenopus laevis* and *Rana pipiens* and from liver cells of *Gallus domesticus* (Wolstenholme and Dawid 1967). Circles of this type appear to characterise mitochondrial DNA.

The question of how the DNA and histone are arranged in the eukaryote chromosome is largely unsolved. We have, however, a few clues:

(i) The length of the DNA molecules extracted from the nuclei of animals makes it unlikely that DNA is present in the form of a single molecule. Applying the Cairn's technique of labelling the DNA fibres in the nuclei of the chinese hamster autoradiographically, Huberman and Riggs (1966) found the largest visible autoradiogram to be 1.1 mm long while 50% of the autoradiograms could be accounted for by lengths equal to or greater than 0.5 mm.

If such autoradiograms are in fact close to the true length of chromosomal DNA molecules then clearly there must be many such molecules per chromosome. In *Lilium,* for example, where there is some 53×10^{-12} gm of DNA per haploid complement, Taylor (1957) has calculated that this DNA must total a length of some $1.5 \times 10^7 \, \mu = 15$ metres. Certainly there are many independent replication points per ~~chromosome~~ (see p. 22) and this too would be most simply explained in terms of independently replicating DNA molecules.

(ii) There is now no doubt that the individual chromatids of many mitotic and meiotic chromosomes are longitudinally subdivisible. Trosko and Brewen (1966) using a combination of hypotonicity, chelation and air-drying on unfixed Chinese hamster chromosomes have shown convincingly that each chromatid is subdivided into at least two half chromatids. The appearance of air-dried metaphase chromosomes of *Vicia faba* after trypsin digestion (Trosko and Wolff 1965) likewise indicates that each chromatid consists of two axes (see Fig. 44). Taylor (1966) has claimed that such subunits are not seen in the living state or in well fixed preparations. In fact both these states do reveal clear half-chromatids (see for example Bajer 1965 and Giménez-Martín et al. 1963 respectively). At least three possible interpretations of these half-chromatids can be offered:

(i) Each half-chromatid is itself a folded and coiled DNA duplex,

(ii) the half-chromatids are themselves the half-helices of a DNA duplex so that the chromatid is a single Watson-Crick unit, or

(iii) each half-chromatid is itself a combination of at least two DNA duplexes.

Taylor (1966) favours (ii) and this is also the conclusion reached by Callan and Gall from their studies on lampbrush chromosomes. Indeed Callan (1967) has stated that the dimensions of the thickness of the fibre between successive chromomeres (ca. 50 Å) and of the axial fibre of the lateral loop (ca. 30 Å) as determined from dried surface film preparations of lampbrush chromosomes (Miller 1965) *"exclude the possibility that the DNA component of these chromosomes is multistranded".* Miller himself is not, however, quite so forthright. What in fact he says is that the diameter of the lateral loop axis (30–50 Å) is *"similar to that obtained by*

other methods for single double helix DNA (WILKINS 1956) *or DNP molecules* (ZUBAY and DOTY 1959) *or for two double helices of DNA precipitated
together* (HALL and CAVALIERI 1961)". CALLAN has also leaned heavily on
GALL's (1963) study of the breakage kinetics of DNase on lampbrush chromosomes. Assuming that in a given region of the chromosome there are (n)
longitudinal subunits which are attacked independently and with equal
probability by the enzyme DNase and assuming also that a visible break
occurs only when all sub-units have been digested at approximately the
same level, then the number of visible breaks (b), will be given as a function of time (t), by the equation

$$b = k_1 t^n$$

in which (k_1) is a proportionality constant. This may be rewritten in
the form

$$\log b = n/\log t + k_2$$

so that a plot of $\log b$ against $\log t$ should yield a straight line of slope (n).

Data on the fragmentation of a group of giant loops near the middle
of chromosome 10 in *Triturus* gave $n_{23} = 2.6 \pm 0.2$. By contrast interchromomeric breakage gave $n_{18} = 4.8 \pm 0.4$. The ratio $4.8/2.6 = 1.8$ confirms that
there are twice as many subunits in the interchromomeric region as in the
loop axis. GALL has also suggested that the true number of subunits is in
fact 4 and 2 respectively. Notice, however, that these values fall respectively 3 and 2 standard errors away from the experimentally determined
means. This evidence is thus by no means as unequivocal as some have
argued it to be (see TAYLOR 1966, CALLAN 1967).

Conventional electron microscopy has been of limited value in inter-
preting chromosome fine structure. This stems partly from the fact that
the method is inadequate for recognizing order which is evident with the
light microscope. Thus electron micrographs of mitotic chromosomes fail
to show the helical order visible with the light microscope. This applies
also to the inability to distinguish centric constrictions in EM sections
(contra JOKELAINEN 1967). Particularly pertinent to this question are phase
contrast studies on mitosis in endosperm of *Haemanthus* (BAJER 1966). These
reveal in a most convincing manner that the coiling of the chromatid, the
lack of coiling at the centromere and the bipartite nature of the anaphase
chromatid are intrinsic properties of chromosomes. In the second place the
dangers of misinterpretation need continually to be borne in mind in EM
studies. This is admirably illustrated by the helical structures described
in the nuclei of *Amoeba proteus* and *Pelomyxa carolinensis* (PAPPAS 1956,
PAPPAS and BRANDT 1958). PAPPAS and BRANDT (*loc. cit.*) first suggested that
these helices might be the DNA-containing components of the nucleus, a
suggestion adopted by TAYLOR (1963) who, in fact, specifically interpreted
them as G_2 chromosomes. These helices label consistently with ^{3}H-RNA
precursors but not with ^{3}H-TdR (WOLSTENHOLME 1966, STEVENS 1967). The
study of STEVENS, in particular, has given good grounds for suggesting that
the helices either represent structures by which, nascent or incomplete

ribosomes are mobilized for transport to the cytoplasm or else that they are the machinery for transporting m-RNA.

More recently GALL (1963) has successfully modified the technique of KLEINSCHMIDT et al. (1959) to spread nuclei of animal cells on an air-water interface. Surface films of nuclei so spread can be picked up on carbon

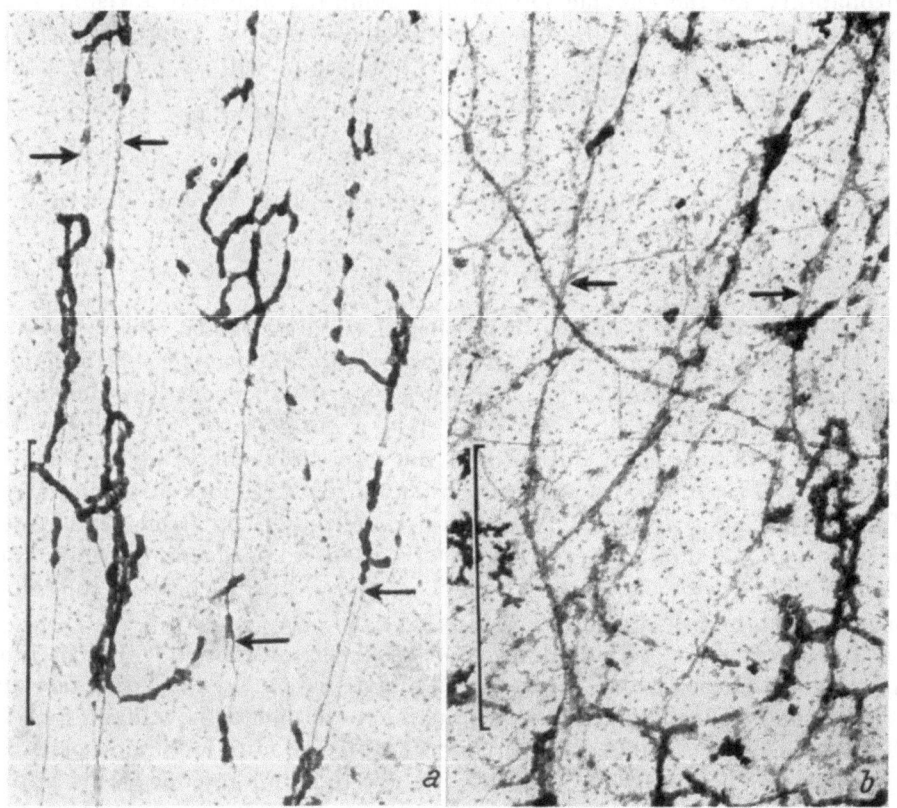

Fig. 43. The structure of air-dried 200 Å fibrils from the nuclei of salamander erythrocytes following spreading on water, fixation in absolute-alcohol and digestion with trypsin (0.5 mg/ml in water adjusted to pH 6.8) for 30 min at 37° C. The arrows mark regions where more than one strand of DNA is left from the original 200 Å thread. (After RIS 1966.)

coated grids and fixed. When dried with ANDERSON's critical point method or simply air-dried from amyl acetate the fine structure of chromosomes is well preserved (WOLFE and JOHN 1965).

The one consistent statement that can so far be made about chromosome organization as inferred from the electron microscopy of both sections and surface films is that the chromosome is built up out of fibrils. The diameters of these have been variously reported from 30 to 500 Å though most commonly they lie in the 100–200 Å range. No definite length can be assigned to them. WOLFE and GRIM (1967) have claimed that the 250 Å fibres most commonly seen in surface-spread preparations probably correspond to the 100 Å fibrils described from EM studies of sectioned nuclei,

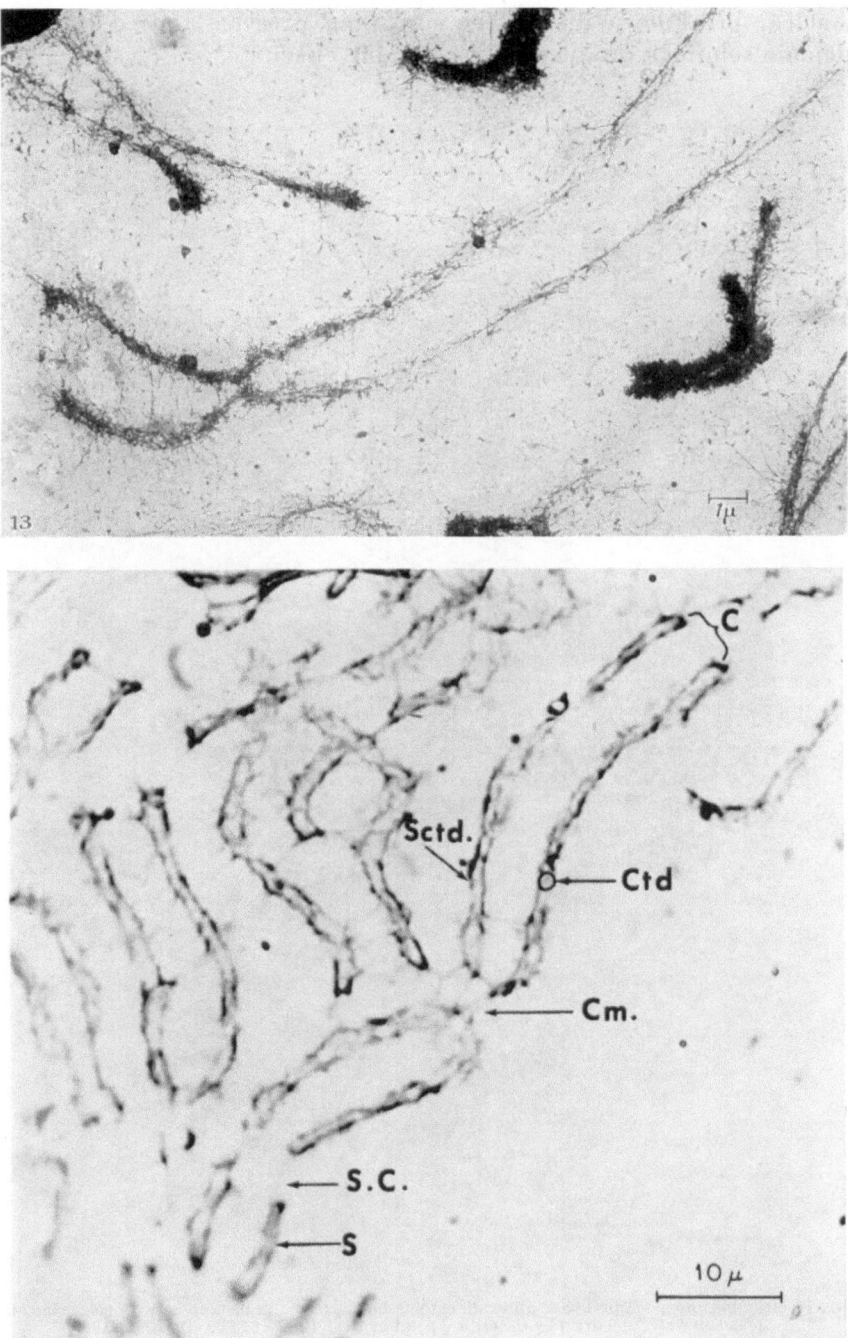

Fig. 44. A comparison of the structure of mitotic chromosomes as seen (a) with the electron microscope after surface spreading and drying with the critical point method (Human c-metaphase, ca. × 5,300, after GALL 1966) and (b) with the light microscope after trypsin treatment of an isolated c-metaphase M-chromosome; note the chromosome (c) is seen with each of its chromatids (Ctd.) divided into two half-chromatids (Sctd.); the satellite (S), secondary constriction (Sc) and centromere (Cm) are all recognisable. (Vicia faba, Feulgen-stained ca. × 1,800, after TROSKO and WOLFF 1965.)

the difference in diameter depending upon a change which occurs early in nuclear breakdown during the spreading process. Indeed lysing with hypotonic solutions produces a comparable effect.

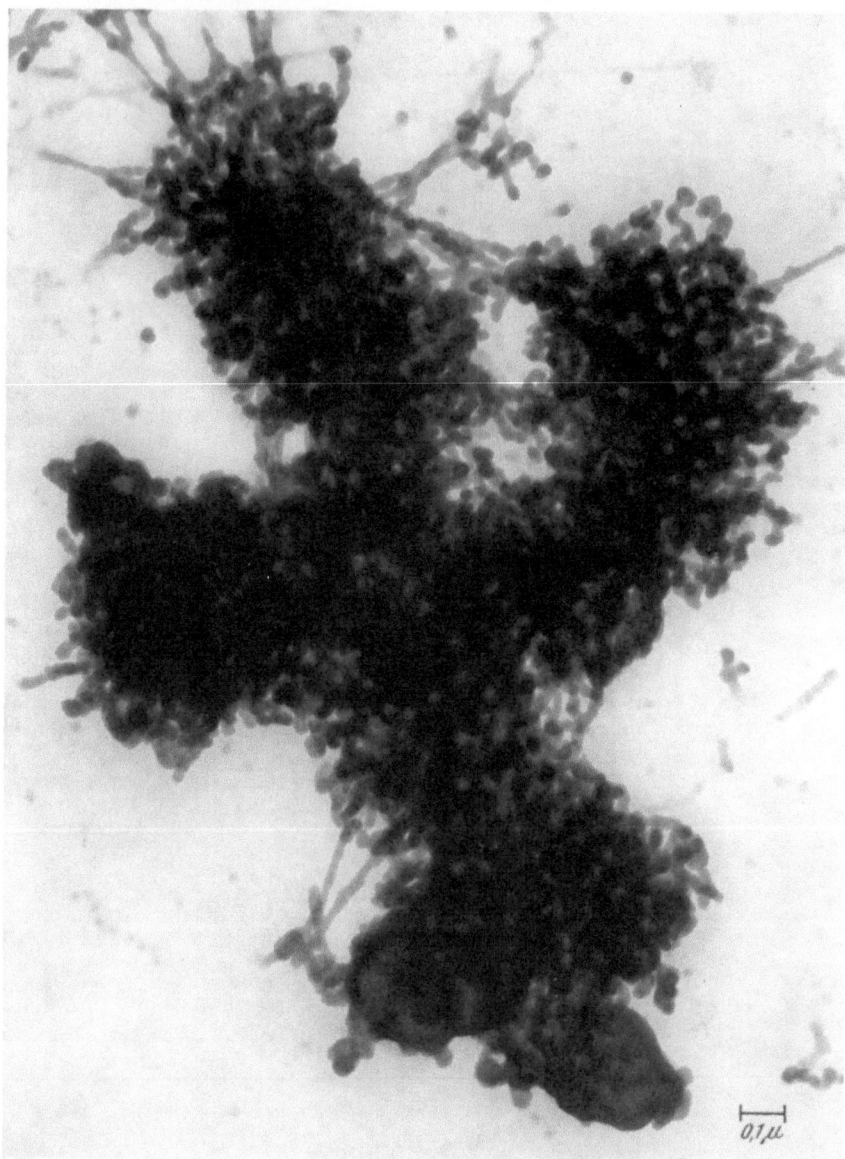

Fig. 45. Electron microscope structure of a human chromosome arrested in metaphase with colchicine and then spread on water and dried by the critical point method. (Ca. × 61,000 after GALL 1966.)

A number of observations demonstrate that these fibrils represent the DNA-histone complex of the chromosomes. For example BASTIA and SWAMINATHAN (1967) find that DNase digestion following spreading does not result in the disruption of microfibril continuity. Digestion with trypsin results in the uncoiling of microfibrils and leads to a reduction of their diameter to 100–40 Å but again does not disrupt the continuity of the fibrils.

Du Praw (1965, 1966) has argued that the 230 Å fibrils he has observed following spreading each consists of a single Watson-Crick DNA duplex held in the form of a regular secondary helix by a proteinaceous sheath. He also believes that chromosome organization results from the folding and coiling of one such a fibril.

There is, in fact, no evidence at all for the first of these assumptions. The naive claim that forks observed in single fibrils represent replication sites (Du Praw 1965) is contradicted by the finding of equivalent forks in non-replicating nuclei (Ris 1966). The most authorative recent study is undoubtedly that of Ris (1967). If erythrocyte nuclei of the salamander *Triturus viridescens* are spread by the monolayer technique, the fibrils obtained are 200–250 Å in diameter. If spreading is carried out in 5×10^{-3} M sodium citrate or 10^{-3} M EDTA, instead of water, the 250 Å fibrils are transformed into strands of 80–100 Å. Moreover, buffers commonly used in E.M. fixatives (phosphate, veronal acetate) act like chelating agents (Ris 1968). If after spreading on water, but before fixation, the fibrils are treated briefly with 5×10^{-3} M sodium citrate the 250 Å fibrils are seen to consist of two parallel strands. When the 100 Å fibrils obtained through the use of chelating agents are digested with pronase, the thin strand that remains is 25 Å in diameter and is sensitive to DNase. The DNA of λ-phage has an identical character. Ris therefore concludes that the 100 Å fibril contains a single DNA dimer combined with protein. If this is so then the 250 Å fibrils contain two DNA dimers. Again, when the 200 Å fibrils, obtained by spreading nucleated erythrocytes on water, are digested with trypsin, very long thin strands remain which can be removed specifically with DNase (Ris 1966). Significantly more than one strand is frequently left from the original 200 Å thread after trypsin treatment (Fig. 43). Ris, therefore, believes that two DNA molecules are bound together by histone to form a fibril and that histone and divalent cations fold the DNA into supercoils to produce the 100–200 Å thick fibres.

The second of Du Praw's assumptions fails to accommodate several relevant facts. In the first place there is no doubt that the spreading technique generates considerable surface tension forces. These tend to disrupt fibril continuity and lead to considerable lateral displacement. This, in turn, produces an over-complicated, even confused, picture of chromosome organization. Compare, for example Figs. 44 a and 45. The former matches favourably with the illustrations of Trosko and Brewen (1966) and Trosko and Wolff (1965, see Fig. 44 b). By contrast Fig. 45 is difficult to reconcile with Fig. 44 a, giving every indication of breakage and displacement of microfibrils. And it is upon pictures of this type that Du Praw has based his folded fibre concept. Secondly, the structure of the cross-shaped diakinesis bivalents obtained by Wolfe and John (1965) is incompatible with DuPraw's hypothesis *.

* See note 10 Appendix.

The balance of evidence, as it stands at the moment, appears to us to favour a model in which each chromatid can be composed of two DNA duplexes. Such a model explains:

(i) The occurrence of departures from the semiconservative distribution of labelled DNA (see p. 19).

(ii) The visible doubleness of chromatids at the light microscope level in both fixed and living cells.

(iii) The occurrence of half-chromatid aberrations (see PEACOCK 1965).

(iv) The mechanical independence of sub-chromatid units in certain coccids (HUGHES-SCHRADER 1940).

In *Vicia faba,* in particular, there is a strong case on the basis of evidence from the light microscope, X-irradiation, autoradiographic analysis (PEACOCK 1965) and trypsin digestion (TROSKO and WOLFF 1965) for the assumption of at least two DNA duplexes per chromatid. This pattern of organization would also accommodate the fact that the chromosomes of *Vicia faba* (EVANS and SAVAGE 1963, WOLFF and LUIPPOLD 1964), like those of Chinese hamster (MONESI, CRIPPA, and ZITO-BIGNAMI 1967), respond to irradiation as double units at the beginning of S and before the bulk of DNA synthesis has occurred. If the chromosome is uninemic and DNA is the unit of breakage this finding can only be accommodated by the separation of the two halves of the DNA duplex (EVANS and SAVAGE 1963). Were this to occur, a proportion of the cells in an asynchronous rapidly growing cell population would be expected to have all or most of their DNA in the single stranded state. Following acridine orange staining, double stranded nucleic acids fluoresce yellow green in UV whereas single stranded nucleic acids fluoresce red. HEDDLE and TROSKO (1966) using acridine orange staining failed to detect cells with single stranded nucleic acid. This too suggests that the chromosome is at least two stranded before S and that the strands are not half DNA duplexes.

Concluding Statement

Thirty and even forty years ago it seemed to many that there was nothing more to be discovered about the genotype at the chromosome level and that genetics should concentrate on the gene (DARLINGTON 1969). However, recent advances involving the combined techniques of experimental breeding, cell biology and biochemistry have so increased our knowledge of the gene that our previous and present ignorance of the chromosome is clearly exposed. Thus, it is now evident that the simple picture of a chromosome as a linear array of unifunctional and ultimately-independent units is quite inadequate for a proper understanding of either genetic or epigenetic systems. Indeed, as we have pointed out else where (LEWIS and JOHN 1963), even the experimental breeder can no longer say that "it is not the shapes and sizes of chromosomes which are important but the genes contained in them" (MORGAN, BRIDGES, and STURTEVANT) and "so long as the sum total of these remains the same, or nearly so, it seems immaterial whether they be grouped to form few or many aggregates" (WILSON). The insufficiency of this view is, however, greater in relation to heredity and evolution than to development and differentiation.

There are, of course, areas and levels of enquiry where to study the chromosome is to study the gene also. This is true, for example, in relation to puff activity in polytene chromosomes and the extension of lampbrush loops. Further, in regard to the specificities of molecular templates the general rules in regard to replication are the same as those for transcription. Here, therefore, we see dramatic support for a prophecy made by BERNARD almost a century ago that "the synthetic act by which the organism maintains itself is, at the bottom, of the same nature as that by which it repairs itself after mutilation, or by which it multiplies and reproduces itself. Organic synthesis, generation, regeneration, reintegration, the healing of a wound, are but different aspects of a single phenomenon". That phenomenon is, of course, heredity which is the transmission of stable, but mutable, self-replicating material from cell to cell in division and fusion (Genesis). But it is revealed not only in terms of this material but in the reappearance in successive generations of like forms and like metabolism (Epigenesis). It is for this reason that "everything of value in biology will gradually take its place in genetics" (DARLINGTON 1951).

In fact, there are parallels between the properties of genetic and epigenetic systems. For example, DARLINGTON (1958) has formulated a number of principles and paradoxes in relation to the former and has pointed out that all the components of the system are related to more than one function and the functions to more than one component. At first sight, this principle of integration appears to be in marked contrast to the unifunctional hypothesis of gene action which underlies discussions of epigenesis. But we must remember that competitive efficiency, differential reproduction and, hence, selective survival are features of whole organisms. It is the peripheral exo- and endo-phenotypes which matter in this regard. What is more, not only do the units of function, mutation and replication themselves differ but they are all far removed from the units of breeding and selection. Thus, even in viral and bacterial systems, while repression, induction and transcription are co-ordinated only for blocks of adjacent structural genes, sequential replication is organized on a whole "chromosome" basis. The organization of recombination, on the other hand, may be more closely related to that of transcription.

It has been suggested that the bands of polytene chromosomes or, at least, their constituent, laterally-multiplied chromomeres, may represent regions which can be separately controlled in respect of both transcription and somatic replication (see p. 89). But it would appear that even the pattern of replication can be differential as between somatic cells within the individual. In fact, integration is nowhere more obvious than in the dual and, in some ways, alternative, auto- and hetero-synthetic activities of the genetic material itself, in the hierarchical nature of gene expression and in the interactions observed between the products of both allelic and non-allelic genes.

9*

What is more, integration may embrace components of both the genetic and the epigenetic system. This is evident in and indicated by, for example, the variable and often undefined use of the term super gene. Thus, at one extreme, it is used in relation to a meiotic and, hence, mechanical relationship between particular alleles of a group of non-homologous loci (Darlington and Mather 1949). In this sense the super gene is embraced by the genetic definition of a differential segment (see Lewis and John 1963) as one which occupies a much shorter fraction of the linkage map than that expected on the basis of physical length and endowment of mendelian mutations. It may even map as a point. In extreme cases, a differential segment may cover an entire chromosome, include parts of many chromosomes, embrace groups of chromosomes and even incorporate a whole haploid complement. Structural and numerical relations as well as the genotypically controlled aspects of pairing, crossing-over and disjunction are involved in the creation and maintenance of these segments. And their genetic definition makes no reference to the causal aspects of their differential (versus pairing) behaviour.

Thus, differential segments may or may not be homologous, they may or may not pair and crossing over may or may not occur in them (cf. Ohno 1967). However, except for the fact that they constitute co-adapted, relationally balanced gene complexes, there is no necessary or direct functional relationship between the linked loci they include. But, clearly the use of the term is not prohibited by such a relationship and it has, in fact, been applied to "complex loci" of various kinds. For example, it has been suggested that the S region which is involved in the control of distyly in primroses is a complex region with three or perhaps even seven tightly linked "sub-genes" (Dowrick 1956). In this case, linkage is afforded by proximity and the procentric location of the region may be important in some species. All the constitutent "cistrons" of such a region function in relation to a particular aspect of the (endo)phenotype and thus show a more obvious and direct co-adaptation than that inferred in relation to differential segments in general. Similar considerations apply in relation to polygenic aggregates and to the "sex genes" borne on the differential segments of X and Y chromosomes. But although these gene complexes are epigenetically important in relation to specific developmental pathways, there is no reason to suppose that their co-ordinated activities, in themselves, require linkage. In other words, their functions, and the integration of these functions, are not likely to depend on their physical proximity and the loci involved are not expected to show position effects. This means that their linkage reflects a genetic and not an epigenetic requirement.

However, in many cases, the evolution of sex chromosome systems has, itself, imposed an epigenetic requirement which arises owing to the hemizygous state of the differential segment of the X chromosome in the heterogametic sex. This requirement is satisfied by the development of systems of dosage compensation. Two systems of this kind have been revealed which differ not only in their mode of operation but in their consequences. One mechanism, exemplified by *Drosophila,* appears to operate in a piece-meal manner so that various loci on the X are compensated independently of others but both members of an allelic pair continue to function. In mammals, on the other hand, one member of each such pair is inactivated in a given cell within females. But the choice seems to be random as between cells. What is more, this process involves a kind of block reaction which appears to achieve co-ordinate compensation for the inactivated alleles of a particular cell all belong to the same chromosome. There are indications that, while the basic action of the differential genes on the mammalian X is not affected by their linkage, spatial relations are important in regard to their compensation (LEWIS and JOHN 1968). Indeed, altered linkage relations can impose compensation on autosomal regions which normally neither require nor show it. Thus, while proximity may be favoured initially for genetic reasons, epigenetic adjustments may subsequently be required. Consequently, it is not always easy to assess the adaptive utility of processes which are relevant to both, such as pycnotization or those structural rearrangements which are associated with position effects.

The term super-gene has been applied also to gene clusters the members of which show an even greater functional relationship, one aspect of which depends on position. Thus not only are the structural genes within an operon functional in the same metabolic pathway, but they are subject to co-ordinate induction or repression at the level of transcription. This co-ordinate control depends on linkage to an operator region, mutations of which can show cis-trans position effects when combined with changes elsewhere in the operon. In this respect they contradict the complementation test frequently used in the definition of a cistron. So do certain X-borne mutations in mammals, owing to the mosaic condition created by the system of dosage compensation to which they are subjected in females.

Of course, the limits of the operon can be adjusted, upwards and downwards, by mutation or rearrangement and, provided that they posses similar operator regions, there is no reason why spatially separated blocks could not be subjected to co-ordinate control. Indeed, such a basis for co-ordination may exist in higher organisms, for the classical operon has been revealed only in viral and bacterial systems. However, it is clear that the organization of an operon fulfils an epigenetic requirement rather than a genetic one and, in this sense, it represents the other extreme situation to which the term super-gene has been applied.

The sub-units of the operon have qualitatively different and, hence, complementary functions. But the super-gene concept has been extended to cover multiple structures whose components are functionally similar and, hence, supplementary, e.g. nucleolar organizer and centromeres. Owing to its mechanical function, the latter is a special case but it can be said of it that, not the activity, but the co-ordinated activity of its constituents depends on proximity for otherwise all the mechanical problems associated with dicentricity would obtain.

To some extent these extensions of the polygene concept parallel the altered attitude towards the gene and its gradual re-definition and operational recognition in terms of functional rather than genetical properties.

The gap between the gene and the character which MENDEL revealed, and which BATESON's demonstration of interaction served both to narrow and widen, is gradually being closed. The classical studies on the eye colours of *Ephestia* and *Drosophila* and the petal pigments of plants paved the path for the more recent and neo-classical investigations on gene-mediated RNA and polypeptide synthesis. But although much remains to be discovered in this connection, the principal areas of ignorance lie not so much within defined disciplines but between them. Thus, at the level of the light microscope, much is known regarding the nature and behaviour of chromosomes during division. Our knowledge of the nature of genes and their activities is no less extensive. But chromosome organization remains a subject for little more than speculation and we are no less ignorant of the control mechanisms which integrate epigenesis.

In this monograph we have attempted to review and order some of the investigations which have lighted the umbra of this ignorance. There is, however, a warning we must take. In the introduction attention was drawn to the conflict and confusion which occurred during the early history of genetics—a state of affairs which arose largely from a lack of sympathy between investigators who used different methods to study the same phenomena. Nowadays, an even greater variety of techniques are employed. They are also more intricate and they develop and change more rapidly. Indeed, it is often only in retrospect that the shortcomings and errors of many experimental methods can be appreciated. Of course, it matters little if an isolated fact proves to be false. But if it constitutes an important link in a chain of inference, then, clearly, the validity of the conclusions can stand or fall on the correctness of the fact.

Biochemical cytology provides many examples of mistaken hypotheses which arose largely from too great a faith in the precision of the experiments on which they were based. Thus, in 1930, NEEDHAM and NEEDHAM claimed that the nucleic acid phosphorus content and, hence, the amount of nucleic acid itself, did not increase between fertilization and the pluteus stage in sea-urchins. And they argued that the cytoplasm of these eggs contained a store of DNA which was used in the genesis of the large number of nuclei which were produced during this period.

In subsequent studies, however, the cytoplasm of these eggs and embryos appeared to be Feulgen negative and only traces of DNA, attributable to contamination from nurse cells, could be found in them. As a result of these findings, but still on the assumption that the nucleic acid content remained constant, the RNA → DNA conversion hypothesis was born (see BRACHET 1957). But this hypothesis too was discarded when it became clear that the "facts" on which it was based were obtained using methods of insufficient specificity.

However, at a still later date there were claims for the presence of large amounts of DNA in the cytoplasm of plant cells (CHAYEN and NORRIS 1953, GAHAN, CHAYEN, and SILCOX 1962). But attempts by BRANTON and RUCH (1963) to confirm these findings were unsuccessful. The existence of a DNA store in oocytes was also re-claimed in the nineteen fifties by many investigators (see BRACHET 1957) using a variety of organisms (*e.g.* sea urchins, frog, chick and *Gryllus)* and various techniques (*e.g.* microbiological assay, phosphorous estimation, isotope dilution and colourimetry).

Of these claims, BRACHET (1957) wrote "this reserve of DNA might not really exist; its quantitative importance decreases when the specificity of the techniques used for DNA determination improves". However, there have been more recent claims for extra DNA in oocytes. Some, but not all, of this appears to be chromosomal (see p. 56).

Nowadays two principal methods are employed in studies of DNA values and synthesis, namely, autoradiography and microdensitometry. The resolving power of the former has been discussed by EVANS and SAVAGE (1963) in relation to the mitotic chromosomes of *Vicia faba.* Thus, they calculated that in the case of thymidine tritiated only in the 5-methyl position, having a specific activity of 4.7 c/m mole and a half-life of 12.26 years, over 15% of the newly incorporated thymine must come from the applied solution to give an average of two silver grains in the emulsion overlying a 1 μ length of metaphase chromatid. This conclusion was based on the following assumptions and observations: i) The total metaphase length of the 24 chromatids in *Vicia faba,* under the cytological conditions employed, is 211 μ, ii) the $A + T : G + C$ in this species is about 1.5 : 1, iii) the 4 C DNA value is approximately 3×10^{-11} grams, and iv) 200 disintegrations from tritium are required to activate one silver grain in the emulsion overlying the labelled locus.

It is not surprising, therefore, that small amounts of DNA synthesis may go undetected by this method. Indeed, as we have already indicated, while a period of DNA synthesis during meiotic prophase can be detected autoradiographically in newts, the method is not sufficiently accurate to yield results in lilies (see p. 27).

Quite large errors may be introduced also when autoradiography is used to determine the duration of the phases in the mitotic cycle. Further, while the root apex is frequently treated as a homogeneous tissue, it is actually composed of a number of zones which differ considerably from each other in regard to the duration of the mitotic cycle and its component phases. In fact CLOWES has determined the duration of the mitotic parameters in various parts of the root apex of maize using a combination of autoradiography and metaphase accumulation, on the one hand, and information from microdensitometry, on the other. A sample of the two sets of results are shown in Table 30. While some of the errors intrinsic to the methods employed may contribute to the differences between the two sets of data, those in relation to the quiescent centre in particular are too great to be accommodated in this way. Consequently CLOWES (1968) has suggested that nuclei may enter mitosis before the 4 C level is reached or the mitotic cycle of quiescent centre cells may be held up, not in G_1, but during S itself. It is also possible that radiation damage may follow the use of radio active precursors for high radiation levels are known to stop normal meristematic cells from dividing while cells of the quiescent centre are stimulated in this respect. The role of an intranuclear source of radiation has been discussed also in relation to the incidence of sister chromatid exchange at mitosis.

Further, conflicting reports have appeared regarding the distribution of DNA within the chromosome. For example, it has long been a matter of debate whether the interband regions of polytene chromosomes contain DNA or not. And recent studies using enzymic digestions and fluorescent microscopy indicate that they do (see p. 33).

In fact, the whole history of the nucleic acids is symptomatic of the flux which has characterized this area of cell biology. Thus, under the older name of thymonucleic acid, DNA was supposed to occur only in animal cells while RNA or zymonucleic acid was believed to be a specific constituent of plants. On this basis, the high RNA content of pancreatic cells was regarded as a curiosity much like the presence of haemoglobin

Table 30. *Comparative DNA Values in Two Regions of the Root Apex of Zea mays.*
(Data of CLOWES 1968.)

| Method of Measurement | Percent Cells | | | | | |
| | Quiescent Centre | | | Cap Initials | | |
	2C	2–4C	4C	2C	2–4C	4C
1. Observed photometrically	53	43	4	4	33	62
2. Estimated from pulse labelling (see also Table 3)	87	5	8	0	57	43

in certain bacterial nodules of plants. Later a nuclear/cytoplasmic segregation of these acids was proposed but this idea was soon replaced by the view that, although DNA was confined to he nucleus, RNA occurred in both nucleus and cytoplasm. Today, of course the presence of extra-chromosomal DNA is acknowledged and the nucleus has been revealed as the principal source of cytoplasmic RNA. Likewise, ideas regarding the structure of DNA have crystallized only during the last decade. And while changes in detail have been introduced since the model of WATSON and CRICK was proposed, the finding of single stranded DNA in the infective phase of the phage Φ X 174 certainly came as a considerable surprise. The detailed organization of transfer RNA molecules, on the other hand, still remains a matter for discussion.

Changes in opinion regarding the role of the nucleic acids have been no less dramatic. For example, the coincidence between the disappearance of the RNA-containing nucleolus and the increasing chromaticity of the chromosomes during prophase, like the opposite sequence at telophase, seemed to support the RNA-DNA conversion hypothesis. This view, based originally on studies of nucleic acid content in sea urchin eggs, was held for a long time. But perhaps the most degrading role attributed to DNA was that of an intra-nuclear buffer! In fact, it is sobering to note that as late as 1955, MARSHAK and MARSHAK claimed that DNA was absent from the eggs of sea urchins and, consequently, it could not play the specific genetic role which had been attributed to it. However, the possibility that scrapie in sheep may be infectively transmitted by principles which do not contain nucleic acid may mean that the last word has not been said in this connection.

From these few examples it is abundantly clear that no aspect of the contemporary study of living organisms has changed, and is changing, as much or as quickly as that considered in this monograph. For this reason we have, atypically, not attempted a general synthesis and have frequently refrained from discussing in detail the significance of the investigations described. The pitfalls we have thus avoided can be illustrated by the following situation.

There is evidence to indicate that:

1. Heteropycnotic regions typically show delayed DNA synthesis (see p. 22).

2. Segments which show the greatest delay in respect of mitotic DNA synthesis, contain the most chiasmata at meiosis (see p. 24).

3. "DNA from the heterochromatic B-chromosomes of maize is made up largely or entirely of G-C pairs" (SCHAIK and PITOUT 1966), and

4. the DNA synthesized during meiotic prophase has a higher G-C content than that produced during the principal S phase (see p. 27).

The inference which can be drawn here is clear. However, the above claim in regard to the G-C content of heterochromatin was not confirmed by the original authors when the DNA was extracted from leaves and seedlings rather than endosperm (SCHAIK, PITOUT, and NEITZ 1967). In fact, the studies of RINEHART (1966), like those of SANSLING which he quotes, did not reveal any difference in G-C content between maize with and without B chromosomes. In these studies CsCl centrifugation, heat denaturation and enzymic hydrolysis followed by column chromatography were employed. It will be appreciated that the kind of heterochromatin referred to here is that which OHNO (1967) has, without complete justification, called "once heterochromatin, always heterochromatin" for, as we have seen (p. 65), and as might be expected, there are no base-ratio differences between chromatin in its iso- and hetero-pycnotic states. However, the tissue differences claimed by SCHAIK et al. remain to be explained. What is more, it has been claimed that the heterochromatic regions revealed by cold treatment in *Vicia* and *Trillium* are characterised by high GC/AT ratios (CASPERSSON et al. 1967). This conclusion is based on the discrete, differential fluorescence shown by these regions following quinacrine mustard treatment, the rationale being that alkylating agents might be expected to interact preferentially with guanine. The classical literature on cytological methods is full of technical variations designed to meet needs of particular materials. Much of this modification of method is doubtless superfluous but it is an aspect of investigation which has been barely considered in relation to more modern techniques.

The range and complexity of these techniques are so great and increasing so rapidly that the results they yield are not easily judged, nor are they easily or quickly checked. Consequently, the investigator now has an even greater responsibility not to publish in haste for the reader has no more leisure to repeat than the author has to repent. In this rapidly changing and expanding field we do well, therefore, to heed Earle Stanley Gardner's advice that "the only way you can keep from swallowing the stuff that is handed out to you today, couched in the semblance of irrefutable logic, is to read the fallacies of yesterday, couched in the same irrefutable logic".

Appendix

The basic outline of this monograph was completed early in 1968. We have, therefore, felt it desirable to include some brief notes on recent papers which have a bearing on statements made in the text.

Note 1 (see p. 23) — Issa et al. (1969) have shown that in rabbit morulae a late replicating X can be demonstrated at least two cell generations before sex chromatin can be identified.

Issa, M., C. E. Blank, and G. W. Atherton, 1969: The temporal appearance of sex chromatin and of the late-replicating X-chromosome in blastocysts of the domestic rabbit. Cytogenetics **8**, 219—237.

Note 2 (see p. 21) — Callan and Taylor (1968) consider it unlikely that any DNA synthesis occurring at zygotene/pachytene in *Triturus* would be detected by radioautographic means. They therefore doubt the validity of Wimber and Prensky's conclusion. They also demonstrate that the proximal heterochromatic regions begin and end DNA synthesis later than other regions and it is this that leads to the asynchrony of the DNA replication pattern.

Callan, H. G., and J. H. Taylor, 1968: A radioautographic study of the time course of male meiosis in the newt *Triturus vulgaris*. J. Cell. Sci. **3**, 615—626.

Note 3 (see p. 29) — Smith and King (1968) have confirmed that no SC is formed in the oocytes of females homozygous for C(3)G+. They have also suggested that the major function of the SC is to ensure disjunction of paired homologues irrespective of whether chiasmata are, or are not, present.

Gassner (1960) draws attention to the fact that two types of achiasmatic meiosis occur. In the one (Panorpa and the mantid Bolbe) SC's are present while in the other (*Tipula, Phryne* and *Drosophila*) they are not. If the SC is necessary for chiasma formation it is also clear that chiasmata do not necessarily result from SC formation (see also Moses 1968).

Finally, while the SC dissociates from the bivalent before first metaphase in the chiasmate meioses of mosquitoes, ascomycetes and lilies, in the achiasmate meiosis of Bolbe at least part of the SC remains associated with the bivalent until first anaphase.

Gassner, G., 1969: Synaptinemal complexes in the achiasmatic spermatogenesis of *Bolbe nigra* Giglio-Tos (Mantoidea). Chromosoma (Berlin) **26**, 22—34.
Moses, M. J., 1968: Synaptinemal complex. Ann. Rev. Genetics **2**, 363—412.
Smith, P. A., and R. C. King, 1968: Genetic control of synaptinemal complexes in *Drosophila Melanogaster*. Genetics **60**, 335—351.

Note 4 (see p. 30) — In *Drosophila* oogenesis two of the sixteen cystocytes in each ovariole are pro-oocytes while the remainder develop into nurse cells. Subsequently, however, one of these pro-oocytes also reverts to a nurse cell and this nurse cell contains SC material (KOCH, SMITH and KING 1967).

MOSES (1968) points out that if the development of the oocyte in mosquitoes is comparable in character this might explain the presence of PC's in mosquito nurse cells.

KOCH, E. A., P. A. SMITH, and R. C. KING, 1967: The division and differentiation of *Drosophila* cystocytes. J. Morph. 121, 55—70.
MOSES, M. J., 1968: Synaptinemal complex. Ann. Rev. Genetics 2, 363—412.

Note 5 (see p. 37) — CROUSE and KEYL (1968) have shown that the extra DNA in the most proximal DNA puff of chromosome II in *Sciara coprophila* arises by additional rounds of replication. The extra DNA thus replicates in the same way as structural DNA.

CROUSE, H. V., and H.-G. KEYL, 1968: Extra replications in the "DNA-puffs" of *Sciara coprophila*. Chromosoma (Berlin) 25, 357—364.

Note 6 (see p. 52) — The RNAs synthesized in pachytene nuclei of chinese hamster cells are mostly DNA-like, heterogeneous and contain little ribosomal precursor material. Autoradiography reveals that these specialized RNAs are synthesized on extra-nucleolar chromatin and apparently break down rapidly. MURAMATSU et al. (1969) theorize that these DNA-like RNAs may have some specific role in the process of meiosis since this is the only system in which this particular type of RNA could be obtained in pure form.

MURAMATSU, M., T. UTAKOJI, and H. SUGANO, 1969: Rapidly labelled nuclear RNA in chinese hamster cells. Exp. Cell Res. 53, 278—283.

Note 7 (see p. 57) — The situation in dytiscids has recently been explored in detail by combined centrifugation and molecular hybridisation techniques (GALL et al. 1969). In both *Dytiscus marginalis* and *Colymbetes fuscus* a high density satellite DNA is found in somatic cells and in sperm. Hybridisation studies show that this DNA is complementary to r-RNA. In oogonia and oocytes the region of high density DNA is replicated extrachromosomally and appears cytologically as a conspicuous chromatin body (Giardina's body).

Ring-shaped multiple nucleoli with DNA-axes occur also in the growing oocytes of *Gryllus domesticus* (KUNZ 1969). The axes of these structures arise from compact DNA-bodies present in oogonia and young oocytes. This system too appears to represent an amplification system for nucleolar genes.

GALL, J. G., H. C. MACGREGOR, and M. E. KIDSTON, 1969: Gene amplification in the oocytes of dytiscid water beetles. Chromosoma (Berlin) 26, 169—187.
KUNZ, W., 1969: Die Entstehung multipler Oocytennukleolen aus akzessorischen DNS-Körpern bei *Gryllus domesticus*. Chromosoma (Berlin) 26, 41—75.

Note 8 (see p. 65) — DNA isolated from XO, XX, XY, XXY and XYY karyotypes of *Drosophila melanogaster* all show a mean GC content of 39.5% despite differences in the relative amount of constitutive heterochromatin (Perreault et al. 1968). That condensed and diffuse chromatin do not differ in histone pattern either has been shown by the use of acrylamide gel electrophoresis (Comings 1967). This has been demonstrated for three kinds of cell models. First, heterochromatin versus euchromatin (human XY versus XXXXY cells; XO versus XYY *Drosophila* and male versus female mealy bugs). Second, metaphase versus interphase cells of a heteroploid human amion cell line and third, mature versus phytohemagluttinin stimulated human lymphocytes.

Comings, D. E., 1967: Histones of genetically active and inactive chromatin. J. Cell Biol. **35**, 699—708.
Perreault, W. J., B. P. Kaufmann, and H. Gay, 1968: Similarity in base composition of heterochromatic and euchromatic DNA in *Drosophila melanogaster*. Genetics **60**, 289—301.

Note 9 (see p. 66) — Whitfield and Perris (1968) find that both inorganic phosphate and phosphoprotein cause complete disappearance of highly condensed chromatin in nuclei isolated from normal thymocytes. Condensed structures reappear, however, when these isolated nuclei are exposed to lysine-rich histone.

Whitfield, J. F., and A. D. Perris, 1968: Dissolution of the condensed chromatin structures of isolated thymocyte nuclei and the disruption of deoxyribonucleo-protein by inorganic phosphate and a phosphoprotein. Exp. Cell Res. **49**, 359—372.

Note 10 (see p. 97) — An extended criticism of the folded fiber model of Du Praw is to be found in Wolfe (1969). Sheldon Wolff (1969) also discusses the much vexed issue of chromosome organization and concludes that the preponderance of cytological evidence indicates that most chromosomes are not single-stranded.

Wolfe, S. L., 1969: The molecular organization of chromosomes. Biological Basis of Medicine, Chapt. **4** (in press).
Wolff, S., 1969: Strandedness of chromosomes. Inter. Rev. Cytol. **25**, 279—296.

Acknowledgments

The authors are grateful to the following sources for permission to use previously published figures:

(1) *Brookhaven Symposia in Biology*
Fig. 4, p. 65, from Stern and Hotta, 1963.

(2) *Chromosoma (Berlin)*
Figs. 6—19, p. 319, vol. 7, from Breuer and Pavan, 1955.
Figs. 13 a and b, p. 93, vol. 15, from Pelling, 1962. — The colour plates for Fig. 26 a (p. 60) were placed at our disposal by Messrs. Carl Zeiss, Oberkochen, West Germany.
Fig. 20, p. 335, vol. 15, from Gabrusewycz-Garcia, 1964.
Figs. 2 a—d, p. 223, vol. 17, from Wolstenholme, 1965.
Fig. 13, p. 229, and Fig. 14, p. 230, vol. 20, from Gall, 1966.
Fig. 1, p. 298, and Figs. 5 a—h, p. 306, vol. 22, from Hennig, 1967.

(3) *Experimental Cell Research*
Fig. 1, p. 610, vol. 30, from Cameron and Prescott, 1963.
Fig. 4, p. 663, vol. 41, from Gledhill, Gledhill, Rigler and Ringertz, 1966.
Figs. 3, 5 and 6, pp. 504 and 505, vol. 42, from Zetterberg, 1966.
Fig. 2, p. 163, vol. 44, from Rees and Evans, 1966.
Fig. 2, p. 326, and Fig. 3, p. 327, vol. 45, from Sheridan and Stern, 1967.

References 109

(4) *Genetics*

Fig. 1, p. 284, vol. 56, from HESS, 1967.

(5) *Japanese Journal of Genetics*

Fig. 1, p. 255, vol. 39, from KUSANAGI, 1964.

(6) *Journal of Cell Biology*

Fig. 1, p. 270, vol. 19, from KONRAD, 1963.
Fig. 13, p. 333, and Figs. 14—17, p. 334, vol. 22, from BLOCH and BRACK, 1964.
Fig. 5 b, p. 130, vol. 26, from TROSKO and WOLFF, 1965.
Fig. 1 and Fig. 14, p. 351, vol. 29, from GABRUSEWYCZ-GARCIA and KLEINFELD, 1966.
Fig. 1, p. 3, and Fig. 2, p. 4, vol. 31, from PRESCOTT, 1966.

(7) *McGraw Hill Book Co.*

Fig. 3, p. 78, in: "The molecular control of cellular activity" (1962), from the article by H. SWIFT.

(8) *Planta*

Fig. 4, p. 86, and Fig. 5, p. 87, vol. 69, from LINSKENS, 1966.

(9) *Prentice Hall Inc.*

Fig. 4.5, p. 49, in: "Actions of chemicals on dividing cells" (1967), by B. A. KIHLMAN.

(10) *Proceedings National Academy of Sciences (U.S.A.)*

Figs. 2 and 3, pp. 899 and 900, vol. 49, from VAN'T HOFF and SPARROW, 1963.
Fig. 1, p. 135, and Fig. 5, p. 138, vol. 50, from GRASSO, WOODARD and SWIFT, 1963.
Figs. 1 and 2, p. 1207, and Figs. 3 and 4, p. 1208, vol. 54, from LITTAU, BURDICK, ALLFREY and MIRSKY, 1965.
Fig. 4 a, p. 859, vol. 56, from DAVIDSON, CRIPPA, KRAMER and MIRSKY, 1966.

(11) *Proceedings Royal Society of London*

Figs. 28 and 29, vol. 164 B, from RIS, 1966.

(12) *Protoplasma*

Fig. 1, p. 224, and Fig. 4, p. 229, vol. 60, from HOTTA and STERN, 1965.

(13) *Springer-Verlag*

Fig. 2, p. 127, from DUSPIVA, and Fig. 2, p. 140, and Fig. 7, p. 150, from SACHSENMEIER, in: "Probleme der Biologischen Reduplikation", 1966.

(14) *The Tissue Culture Association, Inc.*

Fig. 8, p. 91, and Fig. 10, right, p. 93, in: "The Chromosome" (1965), from the article by J. H. FRENSTER.

References

ALFERT, M., 1958: Variations in cytochemical properties of cell nuclei. Exp. Cell Res. Suppl. **6**, 227—235.
ALLFREY, V. G., H. STERN, A. E. MIRSKY, and H. J. SAETREN, 1952: The isolation of cell nuclei in non-aqueous media. J. Gen. Physiol. **35**, 529—557.
AMMERMANN, D. VON, 1965: Cytologische und genetische Untersuchungen an den Ciliaten *Stylonychia mytilus* EHRENBERG. Arch. Protistenk. **108**, 109—152.
ANDERSON, N. G., 1956: Cell division I. A theoretical approach to the primeval mechanism, the initiation of cell division, and chromosomal condensation. Quart. Rev. Biol. **31**, 169—199.
ANSLEY, H. R., 1954: A cytological and cytophotometric study of alternative pathways of meiosis in the house centipede *Scutigera forceps* (RAFINESQUE). Chromosoma (Berlin) **6**, 656—695.

Ansley, H. R., 1958: Histones of mitosis and meiosis in *Loxa flavicolis* (Hemiptera). J. Biophys. Biochem. Cytol. **4**, 59—62.

Astbury, W. T., 1947: X-ray structure of actin. Nature (Lond.) **160**, 388—389.

Avanzi, S., A. Brunori, F. D'Amato, V. N. Ronchi, and G. T. S. Mudnozza, 1963: Occurrence of 2 C (G_1) and 4 C (G_2) nuclei in the radicle meristems of dry seeds in *Triticum durum* Desf. Its implications in studies of chromosome breakage and on developmental processes. Caryologia **16**, 533—558.

Baer, D., 1965: Asynchronous replication of DNA in a heterochromatic set of chromosomes in *Pseudococcus obscurus*. Genetics **52**, 275—285.

Bajer, A., 1957: Cine-micrographic studies on mitosis in endosperm III. The origin of the mitotic spindle. Exp. Cell Res. **13**, 493—502.

— 1965: Sub-chromatid structure of chromosomes in the living state. Chromosoma (Berlin) **17**, 291—302.

— 1966: Morphological aspects of normal and abnormal mitosis. Pp. 90—117 in „Probleme der biologischen Reduplikation". Berlin: Springer-Verlag.

Bammi, R. K., and P. Jura, 1966: Effects of phenethyl alcohol on chromosomes of *Allium cepa*. Exp. Cell Res. **41**, 124—130.

Bastia, D., and M. S. Swaminathan, 1967: Ultrastructure of interphase chromosomes. Exp. Cell Res. **48**, 18—26.

Bauer, H., 1932: Die Histologie des Ovars von *Tipula paludosa* Meig. Z. Wiss. Zool. **143**, 53—76.

Bayreuther, K., 1952: Extra-chromosomale Feulgenpositive Körper (Nucleinkörper) in the Oogenese der Tipuliden. Naturwissenschaften **39**, 71.

Beermann, W., 1952: Chromomerenkonstanz und spezifische Modifikationen der Chromosomenstruktur in der Entwicklung und Organdifferenzierung von *Chironomus tentans*. Chromosoma (Berlin) **5**, 139—198.

— 1967: Gene action at the level of the chromosome. Pp. 179—201 in "Heritage from Mendel". Univ. Wisconsin Press.

Berendes, H. D., and H.-G. Keyl, 1967: Distribution of DNA in heterochromatin and euchromatin of polytene nuclei of *Drosophila hydei*. Genetics **57**, 1—13.

Bergerard, J., 1955: Synthèse de l'acide thymonucléique au cours du cycle mitotique des neuroblasts et des cellules nerveuses embryonnaires d'un insecte *Clitumnus extradentatus* Br. (Phasmidae). C. R. Acad. Sci. (Paris) **240**, 564—567.

Bier, K., W. Kunz, and D. Ribbert, 1969: Insect oogenesis with and without lampbrush chromosomes. Pp. 107—115 in "Chromosomes Today", vol. 2. Edinburgh: Oliver and Boyd.

Bloch, D. P., 1962: Histone synthesis in non-replication chromosomes. J. Histochem. Cytochem. **10**, 137—144.

— 1963: The histones: Syntheses, transitions and functions. Pp. 205—221 in "The cell in Mitosis". London: Academic Press.

— and G. C. Godman, 1955: A microphotometric study of the syntheses of desoxyribonucleic acid and nuclear histone. J. Biophys. Biochem. Cytol. **1**, 17—28.

— and H. Y. C. Hew, 1960: Changes in nuclear histones during fertilization and early embryonic development in the pulmonate snail, *Helix aspersa*. J. Biophys. Biochem. Cytol. **8**, 69—81.

— and S. D. Brack, 1964: Evidence for the cytoplasmic synthesis of nuclear histone during spermiogenesis in the grasshopper *Chortophaga viridifasciata* (de Geer). J. Cell Biol. **22**, 327—340.

— R. A. MacQuigg, S. D. Brack, and J. R. Wu, 1967: The syntheses of desoxyribonucleic acid and histone in the onion root meristem. J. Cell Biol. **33**, 451—467.

Boivin, A. R., R. Vendrely, et C. Vendrely, 1948: L'acide deoxyribonucléique du noyau cellulaire, dépositaire des caractères héréditaires: arguments d'ordre analytique. C. R. Acad. Sci. (Paris) **226**, 1061—1063.

Bonner, J., 1965: The molecular biology of development. University Press: Oxford.

Bonner, J., R. C. Huang, and R. V. Gilden, 1963: Chromosomally directed protein synthesis. Proc. nat. Acad. Sci. U.S. **50**, 893—900.

Brachet, J., 1957: Biochemical Cytology. New York: Academic Press.

Branton, D., and F. Ruch, 1964: The localization of DNA and ultraviolet absorptive granules in *Vicia faba* root tips. Exp. Cell Res. **36**, 285—296.

Breuer, M. E., and C. Pavan, 1955: Behaviour of polytene chromosomes of *Rhyncosciara angelae* at different stages of larval development. Chromosoma (Berlin) **7**, 371—386.

BRINKLEY, B. R., and E. STUBBLEFIELD. 1966: The fine structure of the kinetochore of a mammalien cell *in vitro*. Chromosoma (Berlin) 19, 28—43.

BROWN, D. D., and J. B. GURDON, 1964: Absence of ribosomal RNA synthesis in the anucleolate mutant of *Xenopus laevis*. Proc. nat. Acad. Sci. U.S. 51, 139-—147.

BRUNISH, R., D. FAIRLEY, and J. M. LUCK, 1951: Composition of histone prepared from rat liver deoxypentosenucleoprotein. Nature (Lond.) 168, 82—83.

BUCHER, O., 1956: Histologie und mikroskopische Anatomie des Menschen. 2nd. edit. Bern: Verlag Huber.

BULLOUGH, W. S., and E. B. LAURENCE, 1966: The diurnal cycle in epidermal mitotic division and its relation to chalone and adrenalin. Exp. Cell Res. 43, 343—350.

BURTON, D. W., 1968: Initial changes in the deoxyribonucleoprotein complexes of kangaroo lymphocytes stimulated with phytohaemagglutinin. Exp. Cell Res. 49, 300—304.

BUSCH, H., W. C. STARBUCK, E. J. SINGH, and T. S. Ro, 1964: Chromosomal proteins. Pp. 51—71 in "The role of chromosomes in development". Symp. 23 Soc. Study Dev. and Growth. New York: Academic Press.

CAHN, R. D., and R. LASHER, 1967: Simultaneous synthesis of DNA and specialized cellular products by differentiating cartilage cells *in vitro*. Proc. nat. Acad. Sci. U.S. 58, 1131—1138.

CAIRNS, J., 1963: The bacterial chromosome and its manner of replication as seen by autoradiography. J. molec. Biol. 6, 208—213.

CALLAN, H. G., 1967: The organization of genetic units in chromosomes. J. Cell Sci. 2, 1—7.

— and L. LLOYD, 1960: Lampbrush chromosomes of crested newts *Triturus cristatus* (LAURENTI). Phil. Trans. B 702, 135—219.

CAMENZIND, R., and R. B. NICKLAS, 1968: The non-random chromosome segregation in spermatocytes of *Gryllotalpa hexadactyla*. Chromosoma (Berlin) 24, 324—335.

CAMERON, I. L., and D. M. PRESCOTT, 1963: RNA and protein metabolism in the maturation of the nucleated chicken erythrocyte. Exp. Cell Res. 30, 609—612.

— and D. S. NACHTWEY, 1967: DNA synthesis in relation to cell division in *Tetrahymena pyriformis*. Exp. Cell Res. 46, 385—395.

CASPERSSON, T., 1956: Cytochemistry of nuclear elements. Pp. 73—103 in "Conference on chromosomes". Zwolle: Tjeenk-Willink.

— S. FARBER, G. E. FOLEY, J. KUDYNOWSKI, E. J. MODEST, E. SIMONSSEN, U. WACH, and L. ZECH, 1967: Chemical differentiation along metaphase chromosomes. Exp. Cell Res. 49, 219—222.

CAVE, M. D., 1967: Chromosomal ^{3}H-lysine incorporation and patterns of deoxyribonucleic acid synthesis in human cells. Exp. Cell Res. 45, 631—637.

CHAYEN, J., and K. P. NORRIS, 1953: Cytoplasmic localization of nucleic acids in plant cells. Nature (Lond.) 171, 472—473.

CHIPCHASE, M. I. H., and M. L. BIRNSTIEL, 1963: On the nature of nucleolar RNA. Proc. nat. Acad. Sci. U.S. 50, 1101—1106.

CLEFFMANN, G., 1968: Regulierung der DNS-Menge im Makronucleus von *Tetrahymena*. Exp. Cell Res. 50, 1953—207.

CLOWES, F. A. L., 1965: The duration of the G_1 phase of the mitotic cycle and its relation to radiosensitivity. New Phytol. 64, 355—359.

— 1968: The DNA content of the quiescent centre and root cap of *Zea mays*. New Phytol. 67, 631—639.

COMINGS, D. E., 1967 a: The duration of replication of the inactive X-chromosome in humans based on the persistence of the heterochromatic sex chromatin body during DNA synthesis. Cytogenetics 6, 20—37.

— 1967 b: Sex chromatin, nuclear size and the cell cycle. Cytogenetics 6, 120—144.

CRIPPA, M., 1966: The rate of ribonucleic acid synthesis during the cell cycle. Exp. Cell Res. 42, 371—375.

DANEHOLT, B., and J.-E. EDSTRÖM, 1967: The content of deoxyribonucleic acid in individual polytene chromosome of *Chironomus tentans*. Cytogenetics 6, 350—356.

DARLINGTON, C. D., 1951: Mendel and the determinants. Pp. 315—332 in "Genetics in the 20th Century". New York: MacMillan.

— 1958: The Evolution of Genetic Systems. 2nd edit. Edinburgh: Oliver and Boyd.

— 1969: What we do not know about chromosomes. Pp. 1—4 in "Chromosomes Today", vol. 2. Edinburgh: Oliver and Boyd.

— and K. MATHER, 1949: The Elements of Genetics. London: Allen and Unwin.

Das, C. C., B. P. Kaufmann, and H. Gay, 1964: Histone-protein transition in *Drosophila melanogaster* II. Changes during early embryonic development. J. Cell Biol. **23**, 423—430.

Das, N. K., 1962: Synthetic capacities of chromosome fragments correlated with their ability to maintain nucleolar material. J. Cell Biol. **15**, 121—130.

Davern, C. I., 1966: Isolation of the DNA of the *E. coli* chromosome in one piece. Proc. nat. Acad. Sci. U.S. **55**, 792—797.

Davidson, D., 1964: RNA synthesis in roots of *Vicia faba.* Exp. Cell Res. **35**, 317—325.

— 1966: The onset of mitosis and DNA synthesis in roots of germinating beans. Amer. J. Bot. **53**, 491—495.

— and N. G. Anderson, 1961: Chromosome coiling: abnormalities induced by polyamines. Exp. Cell Res. **20**, 610—613.

Davidson, E. H., and A. E. Mirsky, 1965: Gene activity in oogenesis. Pp. 77—97 in "Genetic control of differentiation". Brookhaven Sympos. Biol. no. 18.

— M. Crippa, F. R. Kramer, and A. E. Mirsky, 1966: Genomic function during the lampbrush chromosome stage of amphibian oogenesis. Proc. nat. Acad. Sci. U.S. **56**, 856—863.

Dawid, I. B., 1965: Deoxyribonucleic acid in amphibian eggs. J. molec. Biol. **12**, 581—599.

— 1966: Evidence for the mitochondral origin of frog egg cytoplasmic DNA. Proc. nat. Acad. Sci. **56**, 269—276.

Dettlaff, T. A., 1964: Cell divisions, duration of interkinetic states and differentiation in early stages of embryonic development. Advances Morphogenesis **3**, 323—360.

Dewey, W. C., and R. M. Humphrey, 1962: Relative radiosensitivity of different phases in the life cycle of L-P 59 mouse fibroblasts and ascites tumour cells. Radiat. Res. **16**, 503—530.

Dounce, A. L., 1952: The enzymes of isolated nuclei. Exp. Cell Res. Suppl. **2**, 103—119.

Dowrick, V. P. J., 1956: Heterostyly and homostyly in *Primula obconica.* Heredity (Lond.) **10**, 219—236.

Du Praw, E. J., 1965 a: The organization of nuclei and chromosomes in honeybee embryonic cells. Proc. nat. Acad. Sci. U.S. **53**, 161—168.

— 1965 b: Macromolecular organization of nuclei and chromosomes: a folded fibre model based on whole-mount electron microscopy. Nature (Lond.) **206**, 338—343.

— 1966: Evidence for "folded-fibre" organization in human chromosomes. Nature (Lond.) **209**, 577—581.

Duspiva, F., 1966: Enzymatische Aspekte der Mitose. Pp. 120—138 in "Probleme der biologischen Reduplikation". Berlin: Springer-Verlag.

Edström, J.-E., 1964: Chromosomal RNA and other nuclear RNA fractions. Pp. 137—152 in "The role of chromosomes in development". Sympos. 23 Soc. Study Dev. and Growth. New York: Academic Press.

— and W. Beermann, 1962: The base composition of nucleic acids in chromosomes, puffs, nucleoli and cytoplasm of *Chironomus* salivary gland cells. J. Cell Biol. **14**, 371—380.

— J. E., and J. G. Gall, 1963: The base composition of ribonucleic acid in lampbrush chromosomes, nucleoli, nuclear sap and cytoplasm of *Triturus* oocytes. J. Cell Biol. **19**, 279—284.

Elsdale, T. R., M. Fischberg, and S. Smith, 1958: A mutation that reduces nucleolar number in *Xenopus laevis.* Exp. Cell Res. **14**, 642—643.

Erickson, R. O., 1947: Respiration of developing anthers. Nature (Lond.) **159**, 275.

— 1948: Cytological and growth correlations in the flower bud and anther development of *Lilium longiflorum.* Amer. J. Bot. **35**, 729—739.

Evans, G. M., and H. Rees, 1966: The pattern of DNA replication at mitosis in the chromosomes of *Scilla campanulata.* Exp. Cell Res. **44**, 150—160.

Evans, H. J., 1964: Uptake of ³H-thymidine and patterns of DNA replication in nuclei and chromosomes of *Vicia faba.* Exp. Cell Res. **35**, 381—393.

— and J. R. K. Savage, 1963: The relation between DNA synthesis and chromosome structure as resolved by X-ray damage. J. Cell Biol. **18**, 525—540.

— and D. Scott, 1964: Influence of DNA synthesis on the production of chromatid aberrations by X-ray and maleic hydrazide in *Vicia faba.* Genetics **49**, 17—38.

FLAMM, W. G., and M. BIRNSTIEL, 1964: Studies on the metabolism of nuclear proteins. Pp. 230—240 in "The Nucleohistones". Holden-Day: San Francisco.

FORER, A., 1966: Characterization of the mitotic traction system, and evidence that birefringent spindle fibers neither produce nor transmit force for chromosome movement. Chromosoma (Berlin) 19, 44—98.

FRENSTER, J. H., 1965 a: A model of specific de-repression within interphase chromatin. Nature (Lond.) 206, 1269—1270.

— 1965 b: Mechanisms of repression and de-repression within interphase chromatin. Pp. 78—101 in "The chromosome". Pub. Tissue Culture Assoc. Vol. 1.

— V. G. ALLFREY, and A. E. MIRSKY, 1963: Repressed and active chromatin isolated from interphase lymphocytes. Proc. nat. Acad. Sci. U.S. 50, 1026—1032.

GABRUSEWYCZ-GARCIA, N., 1964: Cytological and autoradiographic studies in *Sciara coprophila* salivary gland chromosomes. Chromosoma (Berlin) 15, 312—344.

— and R. G. KLEINFELD, 1966: A study of the nucleolar material in *Sciara coprophila*. J. Cell Biol. 29, 347—359.

GAHAN, P. B., J. CHAYEN, and A. A. SILCOX, 1962: Cytoplasmic localization of DNA in *Allium cepa*. Nature (Lond.) 195, 1115—1116.

GALL, J. G., 1959: Macronuclear duplication in the ciliated protozoan *Euplotes*. J. Biophys. Biochem. Cytol. 5, 295—308.

— 1963 a: Kinetics of deoxyribonuclease action on chromosomes. Nature (Lond.) 198, 36—38.

— 1963 b: Chromosomes and cytodifferentiation. Pp. 119—143 in "Cytodifferentiation and macromolecular synthesis". Sympos. 21 Soc. Study Dev. and Growth. New York: Academic Press.

— 1963 c: Chromosome fibers from an interphase nucleus. Science 139, 120—121.

— 1966: Chromosome fibers studied by a spreading technique. Chromosoma (Berlin) 20, 221—233.

— 1968: Differential synthesis of the genes for ribosomal RNA during amphibian oogenesis. Proc. nat. Acad. Sci. 2, 553—560.

— and H. G. CALLAN, 1962: H^3-uridine incorporation in lampbrush chromosomes. Proc. nat. Acad. Sci. U.S. 48, 562—570.

GALTON, M., K. BENIRSCHKE, and S. OHNO, 1965: Sex chromosomes of the chinchilla: allocycly and duplication sequence in somatic cells and behaviour in meiosis. Chromosoma (Berlin) 16, 668—680.

GASSNER, G., 1968: Synaptinemal complexes: recent findings. J. Cell Biol. 35, 166 a (Abst.).

GIMÉNEZ-MARTIN, G., J. F. LÓPEZ-SÁEZ, and A. GONZÁLEZ-FERNÁNDEZ, 1963: Somatic chromosome structure (observations with the light microscope). Cytologia (Tokyo) 28, 381—389.

GLEDHILL, B. L., M. P. GLEDHILL, R. RIGLER Jr., and N. R. RINGERTZ, 1966: Changes in deoxyribonucleoprotein during spermiogenesis in the bull. Exp. Cell Res. 41, 652—665.

GOROVSKY, M. A., and J. WOODARD, 1967: Histone content of chromosomal loci active and inactive in DNA synthesis. J. Cell Biol. 33, 723—728.

GRAHAM, C. F., 1966: The effect of cell size and DNA content in the cellular regulation of DNA synthesis in haploid and diploid embryos. Exp. Cell Res. 43, 13—19.

— K. ARMS, and J. B. GURDON, 1966: The induction of DNA synthesis by frog egg cytoplasm. Develop. Biol. 14, 349—381.

— and R. MORGAN, 1966: Changes in the cell cycle during early amphibian development. Develop. Biol. 14, 439—460.

GRASSO, J. A., J. W. WOODARD, and H. SWIFT, 1963: Cytochemical studies of nucleic acids and proteins in erythrocyte development. Proc. nat. Acad. Sci. U.S. 50, 134—140.

HÅKANSSON, A., and A. LEVAN, 1957: Endo-duplicational meiosis in *Allium odorum*. Hereditas (Lund) 43, 179—200.

HALL, C. E., and L. F. CAVALIERI, 1961: Four-stranded DNA as determined by electron microscopy. J. Biophys. Biochem. Cytol. 10, 347—351.

HARRIS, H., 1965: The short-lived RNA in the cell nucleus and its possible role in evolution. Pp. 469—477 in "Evolving genes and proteins". New York: Academic Press.

— 1966: Hybrid cells from mouse and man: a study in genetic regulation. Proc. Roy. Soc. (Lond.) B 166, 358—368.

HAYASHI, M., M. N. HAYASHI, and S. SPIEGELMAN, 1964: DNA circularity and the mechanism of strand selection in the generation of genetic messages. Proc. nat. Acad. Sci. U.S. **51**, 351—359.

HAYTER, A. M., and R. RILEY, 1969: The significance of DNA: histone ratios in meiotic synapsis. Pg. 266 in "Chromosomes Today", vol. 2. Edinburgh: Oliver and Boyd.

HEDDLE, J. A., and J. E. TROSKO, 1966: Is the transition from chromosome to chromatid aberrations the result of the formation of single-stranded DNA? Exp. Cell Res. **42**, 171—177.

HENDERSON, S. A., 1964: RNA synthesis during male meiosis and spermiogenesis. Chromosoma (Berlin) **15**, 345—366.

HENNIG, W., 1967: Untersuchungen zur Struktur und Funktion des Lampenbürsten-Y-Chromosoms und der Spermatogenese von *Drosophila*. Chromosoma (Berlin) **22**, 294—357.

HESS, O., 1967 a: Complementation of genetic activity in translocated fragments of the Y-chromosome in *Drosophila hydei*. Genetics **56**, 283—295.

— 1967 b: Morphologische Variabilität der chromosomalen Funktionsstrukturen und der Spermatocytenkerne von *Drosophila*-Arten. Chromosoma (Berlin) **21**, 429—445.

HOTTA, Y., and H. STERN, 1963 a: Synthesis of messenger-like ribonucleic acid and protein during meiosis in isolated cells of *Trillium erectum*. J. Cell Biol. **19**, 45—58.

— — 1963 b: Molecular facets of mitotic regulation I. Synthesis of thymidine kinase. Proc. nat. Acad. Sci. U.S. **49**, 648—654.

— — 1965: Polymerase and kinase activities in relation to RNA synthesis during meiosis. Protoplasma **60**, 218—232.

— and A. BASSEL, 1965: Molecular size and circularity of DNA in cells of mammals and higher plants. Proc. nat. Acad. Sci. U.S. **53**, 357—362.

— M. ITO, and H. STERN, 1966: Synthesis of DNA during meiosis. Proc. nat. Acad. Sci. U.S. **56**, 1184—1191.

— L. G. PARCHMAN, and H. STERN, 1968: Protein synthesis during meiosis. Proc. nat. Acad. Sci. **60**, 575—582.

HOWARD, A., and S. R. PELC, 1953: Synthesis of deoxyribonucleic acid in normal and irradiated cells and its relation to chromosome breakage. Heredity (Lond.), Suppl. **6**, 261—274.

HSU, T. C., W. C. DEWEY, and R. M. HUMPHREY, 1962: Radiosensitivity of the cells of chinese hamster *in vitro* in relation to the cell cycle. Exp. Cell Res. **27**, 441—452.

HUANG, C. C., and J. BONNER, 1965: Histone-bound RNA, a component of native nucleohistone. Proc. nat. Acad. Sci. U.S. **54**, 960—967.

HUANG, C. C., J. BONNER, and K. MURRAY, 1964: Physical and Biological properties of nucleohistones. J. molec. Biol. **8**, 54—64.

HUBERMAN, J. A., and J. G. ATTARDI, 1966: Isolation of metaphase chromosomes from HeLa cells. J. Cell Biol. **31**, 95—105.

— and A. D. RIGGS, 1966: Autoradiography of chromosomal DNA fibers from Chinese hamster cells. Proc. nat. Acad. Sci. U.S. **55**, 599—606.

HUGHES-SCHRADER, S., 1940: The meiotic chromosomes of the male *Llaveiella tacnechina* MORRISON (Coccidae) and the question of the tertiary split. Biol. Bull. **78**, 312—337.

ITO, M., Y. HOTTA, and H. STERN, 1967: Studies of meiosis *in vitro* II. Effect of inhibiting DNA synthesis during meiotic prophase on chromosome structure and behaviour. Develop. Biol. **6**, 54—77.

IZAWA, M., V. G. ALLFREY, and A. E. MIRSKY, 1963 a: The relationship between RNA synthesis and loop structure in lampbrush chromosomes. Proc. nat. Acad. Sci. U.S. **49**, 544—561.

— — — 1963 b: Composition of the nucleus and chromosomes in the lampbrush stage of the newt oocyte. Proc. nat. Acad. Sci. U.S. **50**, 811—817.

JOHN, B., and K. R. LEWIS, 1965: The meiotic system. Protoplasmatologia **VI F 1**. Wien: Springer-Verlag.

— — 1968: The Chromosome Complement. Protoplasmatologia **VI A**. Wien: Springer-Verlag.

JOKELAINEN, P. T., 1967: The ultrastructure and spatial organization of the metaphase kinetochore in mitotic rat cells. J. Ultrastruct. Res. **19**, 19—44.

JONA, R., 1966: La durata del ciclo mitotico nella *Bellavalia romana* determinata per via autoradiografica mediante l'imprego della timidina H³. Caryologia **19**, 429—442.

JONES, K. W., 1965: The role of the nucleolus in the formation of ribosomes. J. Ultrastruct. Res. **13**, 257—262.

KEMP, C. L., 1964: The effect of inhibitors on RNA and protein synthesis on cytological development during meiosis. Chromosoma (Berlin) **15**, 652—665.

KEYL, H.-G., 1965: Duplikationen von Untereinheiten der chromosomalen DNS während der Evolution von *Chironomus thummi*. Chromosoma (Berlin) **17**, 139—180.

— 1966: Increase of DNA in chromosomes. Pp. 99—101 in "Chromosomes Today", vol. 1. Edinburgh: Oliver and Boyd.

— und C. PELLING, 1963: Differentielle DNS-Replikation in den Speicheldrüsen von *Chironomus thummi*. Chromosoma (Berlin) **14**, 347—359.

KIHLMAN, B. A., 1967: Actions of chemicals on dividing cells. New Jersey: Prentice Hall.

KILLANDER, D., and A. ZETTERBERG, 1965: A quantitative cytochemical investigation of the relationship between cell mass and initiation of DNA synthesis in mouse fibroblasts *in vitro*. Exp. Cell Res. **40**, 12—20.

KLEINSCHMIDT, A. und R. K. ZAHN, 1959: Über Desoxyribonucleinsäure-Molekeln in Protein-Mischfilmen. Z. Naturforschg. **14 b**, 770—779.

KLUSS, B. C., 1962: Electron microscopy of the macronucleus of *Euplotes eurystomus*. J. Cell Biol. **13**, 462—465.

KONRAD, C. R., 1963: Protein synthesis and RNA synthesis during mitosis in animal cells. J. Cell Biol. **19**, 267—277.

KUSANAGI, A., 1964 a: Cytological studies on *Luzula* chromosome VI. Migration of the nucleolar RNA to metaphasic chromosomes and spindle. Bot. Mag. Tokyo **77**, 388—392.

— 1964 b: RNA synthetic activity in the mitotic nuclei. Jap. J. Genet. **39**, 254—258.

KUWADA, Y., N. SHINKE, and G. OURA, 1938: Artificial uncoiling of the chromosome spirals as a method of investigation of the chromosome structure. Z. Wiss. Mikr. **55**, 8—16.

LA COUR, L. F., 1963: Ribose nucleic acid and the metaphase chromosome. Exp. Cell Res. **29**, 112—118.

— and J. CHAYEN, 1958: A cyclic staining behaviour of the chromosomes during mitosis and meiosis. Exp. Cell Res. **14**, 402—468.

LACOUR, L. F., and S. R. PELC, 1958: Effect of colchicine on the utilization of labelled thymidine during chromosomal reproduction. Nature (Lond.) **182**, 506—508.

— — 1959: Effect of colchicine on the utilization of thymidine labelled with tritium during chromosomal reproduction. Nature (Lond.) **183**, 1455—1456.

— and B. WELLS, 1967: The loops and ultrastructure of the nucleolus in *Ipheion uniflorum*. Z. Zellforsch. **82**, 25—45.

LEACH, W. M., 1964: Retention of tritiated thymidine in grasshopper neuroblasts. Exp. Cell Res. **35**, 201—204.

LESLIE, I., 1961: Biochemistry of heredity: a general hypothesis. Nature (Lond.) **189**, 260—268.

LEWIS, K. R., and B. JOHN, 1963: Chromosome Marker. London: Churchill.

— — 1968: The chromosomal basis of sex determination. Int. Rev. Cytol. **23**, 277—379.

LEZZI, M., 1965: Die Wirkung von DNase auf isolierte Polytan-Chromosomen. Exp. Cell Res. **39**, 289—292.

— 1967: Cytochemische Untersuchungen an Puffs isolierter Speicheldrüsen-Chromosomen von *Chironomus*. Chromosoma (Berlin) **21**, 89—108.

LIAO, S., and A. H. LIN, 1967: Prostatic nuclear chromatin: an effect of testosterone on the synthesis of ribonucleic acid rich in cytidyl (3′ 5′) guanosine. Proc. nat. Acad. Sci. U.S. **57**, 379—386.

LIAU, M. C., L. S. HNILICA, and R. B. HURLBERT, 1965: Regulation of RNA synthesis in isolated nucleoli by histones and nuclear proteins. Proc. nat. Acad. Sci. U.S. **53**, 626—633.

LIMA-DE-FARIA, A., 1959: Differential uptake of tritiated thymidine into hetero- and euchromatin in *Melanoplus* and *Secale*. J. Biophys. Biochem. Cytol. **6**, 457—466.

— 1962 a: Metabolic DNA in *Tipula oleracea*. Chromosoma (Berlin) **13**, 47—59.

— 1962 b: Progress in tritium autoradiography. Progr. Biophys. **12**, 282—317.

— and M. J. MOSES, 1966: Ultrastructure and cytochemistry of metabolic DNA in *Tipula*. J. Cell Biol. **30**, 177—192.

Lima-de-Faria, A., B. Nilsson, D. Cove, A. Puga, and H. Jaworska, 1968: Tritium labelling and cytochemistry of extra DNA in *Acheta*. Chromosoma **25**, 1—20.

Lin, H. J., J. D. Karkas, and E. C. Chargaff, 1966: Template functions in the enzymic formation of polyribonucleotides II. Metaphase chromosomes as templates in the enzymic synthesis of ribonucleic acid. Proc. nat. Acad. Sci. U.S. **56**, 954—959.

Lindsley, D. L., and E. Novitski, 1958: Localization of the genetic factors responsible for the kinetic activity of X-chromosomes of *Drosophila melanogaster*. Genetics **43**, 790—798.

Linskens, H. F., 1966: Die Änderung des Protein- und Enzym-Musters während der Pollenmeiose und Pollenentwicklung. Physiologische Untersuchungen zur Reifeteilung. Planta (Berlin) **69**, 79—91.

Littau, V. C., C. J. Burdick, V. G. Allfrey, and A. E. Mirsky, 1965: The role of histones in the maintenance of chromatin structure. Proc. nat. Acad. Sci. U.S. **54**, 1204—1212.

MacGregor, H. C., 1965: The role of lampbrush chromosomes in the formation of nucleoli in amphibian oocytes. Quart. J. micr. Sci. **106**, 215—228.

— 1967: Pattern of incorporation of [³H] uridine into RNA of amphibian oocyte nucleoli. J. Cell Sci. **2**, 145—150.

— and H. G. Callan, 1962: The actions of enzymes on lampbrush chromosomes. Quart. J. micr. Sci. **103**, 173—203.

Maio, J. J., and C. L. Schildkraut, 1966: Isolation and properties of mammalian metaphase chromosomes. Fed. Proc. **25**, 707 (Abst.).

Marmur, J., and P. Doty, 1961: Terminal renaturation of DNA. J. molec. Biol. **3**, 585—594.

Marshak, A., and C. Marshak, 1955: Quantitative determination of desoxyribonucleic acid in echinoderm germ cells. Exp. Cell Res. **8**, 126—146.

Martin, P. G., 1966: Variation in the amounts of nucleic acid in the cells of different species of higher plants. Exp. Cell Res. **27**, 84—94.

Matsuura, H., and M. Iwabuchi, 1962: Effect of inorganic salts on cell division II. Production of meiotic abnormalities in PMCs of Paris by NaCl, KCl and $CaCl_2$. J. Fac. Sci. Hokkaido Univ. V **8**, 115—142.

Maul, G. C., and T. H. Hamilton, 1967: The intranuclear localization of two DNA-dependent RNA polymerase activities. Proc. nat. Acad. Sci. U.S. **57**, 1371—1378.

Mazia, D., 1961: Mitosis and the physiology of cell division. Pp. 77—412 in "The Cell", vol. 3. New York: Academic Press.

— and R. T. Hinegardner, 1963: Enzymes of DNA synthesis in nuclei of sea urchin embryos. Proc. nat. Acad. Sci. U.S. **50**, 148—156.

McCarthy, B. J., and B. H. Hoyer, 1964: Identity of DNA and diversity of messenger RNA molecules in normal mouse tissues. Proc. nat. Acad. Sci. U.S. **52**, 915—922.

McGrath, R. A., R. W. Williams, 1967: Interruptions in single strands of the DNA in slime mould and other organisms. Biophys. J. **7**, 309—317.

Meyer, G. F., 1960: The fine structure of spermatocyte nuclei of *Drosophila melanogaster*. Proc. Europ. Reg. Congr. Electron Microscopy.

— 1963: Die Funktionsstrukturen des Y-Chromosoms in den Spermatocytenkernen von *Drosophila hydei, D. neohydei, D. repleta* und einigen anderen *Drosophila*-Arten. Chromosoma (Berlin) **14**, 207—255.

Miller, O. L., 1965: Fine structure of lampbrush chromosomes. Pp. 79—97 in "Genes and chromosomes: structure and function". Nat. Cancer Inst. Monogr. **18**.

Mirsky, A. E., and H. Ris, 1949: Variable and constant components of chromosomes. Nature (Lond.) **163**, 666—667.

— and S. Osawa, 1961: The interphase nucleus. Pp. 677—770 in "The Cell", vol. 2. New York: Academic Press.

Molè-Bajer, J., 1965: Telophase segregation of Chromosomes and amitosis. J. Cell Biol. **25**, 79—94.

Monesi, V., 1965 a: Differential rate of ribonucleic acid synthesis in the autosomes and sex chromosomes during male meiosis in the mouse. Chromosoma (Berlin) **17**, 11—21.

— 1965 b: Synthetic activities during spermatogenesis in the mouse. Exp. Cell Res. **39**, 197—224.

— M. Crippa, and R. Zito-Bignami, 1967: The stage of chromosome duplication in the cell cycle as revealed by X-ray breakage and ³H-thymidine labelling. Chromosoma (Berlin) **21**, 369—386.

Moses, M. J., and J. H. Taylor, 1955: Desoxypentose nucleic acid synthesis during microgametogenesis in *Tradescantia*. Exp. Cell Res. **9**, 474—488.

NEEDHAM, J., and D. M. NEEDHAM, 1930: On phosphorus metabolism in embryonic life. I. Invertebrate eggs. J. Exp. Biol. **7**, 317—348.

NIEHAUS, W. G., and C. P. BARNUM, 1965: Incorporation of radioisotope *in vivo* into ribonucleic acid and histone in a fraction of nuclei preparing for mitosis. Exp. Cell Res. **39**, 435—442.

NILSSON, B., 1966: DNA bodies in the germ line of *Achaeta domesticus* (Orthoptera). Hereditas (Lund) **56**, 396—398.

NUR, U., 1966: Non-replication of heterochromatic chromosomes in a mealy bug, *Planococcus citri* (Coccoidea: Homoptera). Chromosoma (Berlin) **19**, 439—448.

OHNO, S., 1967: Sex Chromosomes and Sex Linked Genes. Berlin: Springer-Verlag.
— W. BEÇAK, and M. L. BEÇAK, 1964: X-autosome ratio and the behaviour pattern of individual X-chromosomes in placental mammals. Chromosma (Berlin) **15**, 14—30.

OKAZAKI, K., and H. HOLTZER, 1966: Myogenesis: fusion, myosin synthesis and the mitotic cycle. Proc. nat. Acad. Sci. U.S. **56**, 1484—1490.

ÖSTERGREN, G., J. MOLÈ-BAJER, and A. BAJER, 1960: An interpretation of transport phenomena at mitosis. Ann. N.Y. Acad. Sci. **90**, 381—408.

— and K. ÖSTERGREN, 1966: Mitosis with undivided Chromosomes III. Inhibition of chromosome reproduction in *Tradescantia* by specific mutations. Pp. 128—130 in "Chromosomes To-day", vol. 1, Edinburgh: Oliver and Boyd.

PAINTER, T. S., 1966: The role of the E-chromosomes in Cecidomyidae. Proc. nat. Acad. Sci. U.S. **56**, 853—855.

— and J. J. BIESELE, 1966: The fine structure of the hypopharyngeal gland cell of the honey bee during development and secretion. Proc. nat. Acad. Sci. U.S. **55**, 1414—1419.

PAPPAS, G. D., 1956: Helical structures in the nucleus of *Amoeba proteus*. J. Biophys. Biochem. Cytol. **2** (Suppl.), 431—434.

— and P. W. BRANDT, 1958: Helical structures in the nuclei of free-living amoebas. Proc. 4th Int. Conf. Elect. Micr. Berlin **2**, 244—246.

PEACOCK, W. J., 1965: Chromosome replication. Pp. 101—123 in "Genes and chromosomes: structure and function". Nat. Cancer Inst. Monogr. **18**.

PELC, S. R., 1964: Labelling of DNA and cell division in so-called non-dividing tissues. J. Cell Biol. **22**, 21—28.

PELLING, C., 1964: Ribonukleinsäure-Synthese der Riesenchromosomen. Autoradiographische Untersuchungen an *Chironomus tentans*. Chromosoma (Berlin) **15**, 71—122.

— 1966: A replicative and synthetic chromosomal unit—the modern concept of the chromomere. Proc. roy. Soc. (Lond.) **B 164**, 279—289.

PERKOWSKA, E., H. C. MACGREGOR, and M. L. BIRNSTIEL, 1968: Gene amplification in the oocyte nucleus of mutant and wild type *Xenopus laevis*. Nature **217**, 649—650.

PERRY, R. P., 1965: The nucleolus and the synthesis of ribosomes. Pp. 325—339 in "Genes and Chromosomes: structure and function". Nat. Cancer Inst. Monogr. **18**.

— and M. ERRERA, 1960: The influence of nucleolar ribonucleic acid metabolism on that of the nucleus and cytoplasm. Pp. 24—29 in "The Cell Nucleus". London: Butterworth.

— A. HELL, and M. ERRERA, 1961: The role of the nucleolus in ribonucleic and protein synthesis. I. Incorporation of cytidine into normal and inactivated HeLa cells. Biochim. Biophys. Acta **49**, 47—57.

POLLISTER, A. W., 1952: Nucleoproteins of the nucleus. Exp. Cell Res. **2** (Suppl.), 59—70.

PLAUT, W., D. NASH, and T. FANNING, 1966: Ordered replication of DNA in polytene chromosomes of *Drosophila melanogaster*. J. molec. Biol. **16**, 85—93.

PRENSKY, W., and H. H. SMITH, 1964: Incorporation of ^{3}H-arginine in chromosomes of *Vicia faba*. Exp. Cell Res. **34**, 525—532.

PRESCOTT, D., 1963: RNA and protein replacement in the nucleus during growth and division in the conservation of components in the chromosome. Pp. 111—128 in "Cell Growth and Cell Division". New York: Academic Press.

— 1964: The normal cell cycle. Pp. 71—97, Chapt. 3, in "Synchrony in Cell Division and Growth". Interscience Pub., New York.

— 1966: The synthesis of total macronuclear protein, histone and DNA during the cell cycle in *Euplotes eurystomus*. J. Cell Biol. **31**, 1—9.

Prescott, D., and M. A. Bender, 1963: Autoradiographic study of chromatid distribution of labelled DNA in two types of mammalian cells *in vitro*. Exp. Cell Res. **29**, 430—442.

Ptashne, M., 1960: The behaviour of strong and weak centromeres at second anaphase of *Drosophila melanogaster*. Genetics **45**, 499—506.

Quastler, H., 1963: The analysis of cell population kinetics. Pp. 18—36 in "Cell Proliferation". Oxford: Blackwell.

Rao, M. V. N., and D. M. Prescott, 1967: Micronuclear RNA synthesis in *Paramoecium caudatum*. J. Cell Biol. **33**, 281—285.

Rasch, E., and J. W. Woodard, 1959: Basic proteins of plant nuclei during normal and pathological growth. J. Biophys. Biochem. Cytol. **6**, 263—276.

Rees, H., and G. M. Evans, 1966: A correlation between the localization of chiasmata and the replication pattern of chromosomal DNA. Exp. Cell Res. **44**, 161—164.

Reid, B. R., and R. D. Cole, 1964: Biosynthesis of a lysine-rich histone in isolated calf thymus nuclei. Proc. nat. Acad. Sci. U.S. **51**, 1044—1050.

Rinehart, K., 1966: Maize DNA composition: analysis of plants with and without B chromosomes. Maize Genet. Coop. News Letter **40**, 56—58.

Ringertz, N. R., J. L. E. Ericson, and O. Nilsson, 1967: Macronuclear chromatin structure in Euplotes. Exp. Cell Res. **48**, 97—117.

Ris, H., 1961: Ultrastructure and molecular organization of genetic systems. Canad. J. Genet. Cytol. **3**, 95—120.

— 1966 a: Fine structure of chromosomes. Proc. roy. Soc. (Lond.) **B 164**, 256—257.

— 1966 b: The organization of chromosomal nucleohistone fibrils. Pp. 339—340 in "Electron Microscopy". Tokyo: Maruzen Co.

— 1967: Ultrastructure of the animal chromosome. Pp. 11—21 in "Regulation of nucleic acid and protein biosynthesis". Elsevier Pub. Co: Amsterdam.

— 1968: Effect of fixation on the dimension of nucleohistone fibers. Amer. Soc. Cell Biol. (Abst.).

Ritossa, F. M., and S. Spiegelmann, 1965: Localization of DNA complementary to ribosomal RNA in the nucleolus organizer region of *Drosophila melanogaster*. Proc. nat. Acad. Sci. U.S. **53**, 737—745.

Robbins, E., and T. W. Borun, 1967: The cytoplasmic synthesis of histones in HeLa cells and its temporal relationship to DNA replication. Proc. nat. Acad. Sci. U.S. **57**, 409—416.

Rodman, T. C., 1967: DNA replication in salivary gland nuclei of *Drosophila melanogaster* at successive larval and prepupal stages. Genetics **55**, 375—386.

Roth, T. F., 1966: Changes in the synaptinemal complex during meiotic prophase in mosquitoe oocytes. Protoplasma **61**, 346—386.

— and M. Ito, 1967: DNA-dependent formation of the synaptinemal complex at meiotic prophase. J. Cell Biol. **35**, 247—255.

— and L. G. Parchman, 1968: Diplotene achiasmatic chromosomes following normal synapsis at pachynema. 25th Ann. EMSA Meeting (Abst.).

Rudkin, G. T., 1963: The structure and function of heterochromatin. Proc. 11th Int. Congr. Genetics **2**, 359—374.

— 1965 a: Nonreplicating DNA in giant chromosomes. Genetics **552**, 470 (Abst.).

— 1965 b: The relative mutabilities of DNA in regions of the X chromosomes of *Drosophila melanogaster*. Genetics **52**, 665—681.

— and S. L. Corlette, 1957: Disproportionate synthesis of DNA in a polytene chromosome region. Proc. nat. Acad. Sci. U.S. **43**, 964—968.

Sachsenmaier, W., 1966: Analyse des Zellcyklus durch Eingriffe in die Makromolekül-Biosynthese. Pp. 139—159 in "Probleme der biologischen Reduplikation". Berlin: Springer-Verlag.

Saint-Amand, G. S., N. G. Anderson, and M. E. Gaulden, 1955: Quoted in "Mitogenesis". Dev. Biol. Conference Series (1956). Univ. Chicago Press 1959.

Sampson, M., A. Katoh, Y. Hotta, and H. Stern, 1963: Metabolically labile deoxyribonucleic acid. Proc. nat. Acad. Sci. U.S. **50**, 459—463.

Schaik, N. van, and M. J. Pitout, 1966: DNA from maize with B chromosomes. Maize Genet. Coop. News Letter **40**, 123—125.

— — and A. W. H. Neitz, 1967: Base ratios of maize DNA. Maize Genet. Coop. News Letter **41**, 167—168.

Schultz, J., 1965: Genes, differentiation and animal development. Pp. 116—147 in "Genetic Control of Differentiation". Brookhaven Symp. Biol. no. 18.

Scott, D., and H. J. Evans. 1964: Influence of the nucleolus on DNA synthesis and mitosis in *Vicia faba*. Exp. Cell Res. **36**, 145—159.

Serra, J. A., 1947: The parallelism between the chemical and the morphological changes in the chromosomes during mitosis and meiosis. Exp. Cell Res. 1 (Suppl.), 111—122.

— 1955: Chemistry of the nucleus. Pp. 1—54 in "Encyclopaedia of Plant Physiology". Berlin: Springer-Verlag.

Shapiro, I. M., and L. H. Levina, 1967: Autoradiographical study on the time of nuclear protein synthesis in human leucocyte blood culture. Exp. Cell Res. **47**, 75—85.

Sheridan, W. F., and H. Stern, 1967: Histones of meiosis. Exp. Cell Res. **45**, 323—335.

Skinner, D. M., 1967: Satellite DNA's in the crabs *Gecarcinus lateralis* and *Cancer pagurus*. Proc. nat. Acad. Sci. U.S. **58**, 103—110.

Solari, J., 1965: Structure of the chromatin in the sea urchin sperm. Proc. nat. Acad. Sci. U.S. **53**, 503—511.

Sotelo, J. R., and O. Trujillo-Cenóz, 1960: Electron microscope study of spermatogenesis. Z. Zellforsch. **51**, 243—277.

Sparvoli, E., H. Gay, and B. P. Kaufmann, 1966: Duration of the mitotic cycle in *Haplopappus gracilis*. Caryologia **19**, 65—71.

Steffensen, D. M., 1953: Induction of chromosome breakage at meiosis by a magnesium deficiency in *Tradescantia*. Proc. nat. Acad. Sci. U.S. **39**, 613—620.

— 1955: Chromosome breakage with a calcium deficiency in *Tradescantia*. Proc. nat. Acad. Sci. U.S. **41**, 155—160.

— 1966: Synthesis of ribosomal RNA during growth and division in *Lilium*. Exp. Cell Res. **44**, 1—12.

Stein, O., and H. Quastler, 1963: The use of tritiated thymidine in the study of tissue activation during germination in *Zea mays*. Amer. J. Bot. **50**, 1006—1011.

Stern, H., and A. E. Mirsky, 1953: Soluble enzymes of nuclei isolated in sucrose and non-aqueous media. J. gen. Physiol. **37**, 177—187.

— and Y. Hotta, 1963: Regulated synthesis of RNA and protein in the control of cell division. Pp. 59—70 in "Meristems and Differentiation". Brookhaven Symp. Biol. 16.

— — 1967: Chromosome behaviour during development of meiotic tissue. Pp. 47—76 in "The Control of Nuclear Activity". New Jersey: Prentice Hall.

Stevens, A. R., 1967: Machinery for exchange across the nuclear envelope. Pp. 189—271 in "The Control of Nuclear Activity". Symp. Soc. Gen. Physiol. New Jersey: Prentice Hall.

Stewart, J. E., and J. Papaconstantinou, 1967: A stabilization of RNA templates in lens cell differentiation. Proc. nat. Acad. Sci. U.S. **58**, 95—102.

Stockdale, F. E., and H. Holtzer, 1961: DNA synthesis and myogenesis. Exp. Cell Res. **24**, 508—520.

Stubblefield, E., 1964: DNA synthesis and chromosomal morphology of Chinese hamster cells cultured in media containing N-deacetyl-N-methyl colchicine (colcemid). Pp. 223—248 in "Symp. Inter. Soc. Cell Biol", vol. 3. New York: Academic Press.

Swanson, C. P., and W. J. Young, 1965: Chromosome reproduction in mitosis and meiosis. Pp. 107—129 in "Reproduction: Molecular, Subcellular and Cellular". 24th Sympos. Soc. Dev. Biol. New York: Academic Press.

Swift, H., 1962: Nucleic acids and cell morphology in dipteran salivary gland cells. Pp. 73—125 in "The Molecular Control of Cellular Activity". London: McGraw Hill.

Tanaka, R., 1965: H³-thymidine autoradiographic studies on the heteropycnosis, heterochromatin and euchromatin in *Spiranthes sinensis*. Bot. Mag. Tokyo **78**, 50—62.

Taylor, E. W., 1965: Control of DNA synthesis in mammalian cells in culture. Exp. Cell Res. **40**, 316—322.

Taylor, J. H., 1957: The time and mode of duplication of chromosomes. Amer. Nat. **91**, 209—221.

— 1958 a: The mode of chromosome duplication in *Crepis capillaris*. Exp. Cell Res. **15**, 350—357.

— 1958 b: Sister chromatid exchanges in tritium labelled chromosomes. Genetics **43**, 515—529.

— 1959: Autoradiographic studies of nucleic acids and proteins during meiosis in *Lilium longiflorum*. Amer. J. Bot. **46**, 477—484.

11a*

TAYLOR, J. H., 1960: A synchronous duplication of chromosomes in cultured cells of Chinese hamster. J. Biophys. Biochem. Cytol. 7. 455—464.
— 1963: The replication and organization of DNA in chromosomes. Pp. 65—111 in "Molecular Genetics". Part. 1. New York: Academic Press.
— 1965: The distribution of tritium labelled DNA among chromosomes during meiosis I. Spermatogenesis in the grasshopper. J. Cell Biol. 25, 57—67.
— 1966: The duplication of chromosomes. Pp. 9—26 in "Probleme der biologischen Reduplikation". Berlin: Springer-Verlag.
— and R. D. McMASTER, 1954: Autoradiographic and microphotometric studies of desoxyribose nucleic acid during microgametogenesis in Lilium longiflorum. Chromosoma (Berlin) 6, 489—521.
— P. S. WOODS, and W. L. HUGHES, 1957: The organization and duplication of chromosomes as revealed by autoradiographic studies using tritium-labelled thymidine. Proc. nat. Acad. Sci. U.S. 43, 122—128.
TERASIMA, T., and M. YASUKAWA, 1966: Synthesis of G_1-protein preceding DNA synthesis in cultured mammalian cells. Exp. Cell Res. 44, 669—672.
TERRA, N. DE, 1967: Macromolecular DNA synthesis in Stentor: regulation by a cytoplasmic initiator. Proc. nat. Acad. Sci. U.S. 57, 607—614.
TOLIVER, A., and E. H. SIMON, 1967: DNA synthesis in 5-Bromouracil tolerant HeLa cells. Exp. Cell Res. 45, 603—617.
TOOZE, J., and H. G. DAVIES, 1963: The occurrence and possible significance of haemoglobin in the chromosomal regions of mature erythrocyte nuclei of the newt Triturus cristatus cristatus. J. Cell Biol. 16, 501—511.
TROSKO, J. E., and S. WOLFF, 1965: Strandedness of Vicia faba chromosomes as revealed by enzyme digestion studies. J. Cell Biol. 26, 125—135.
— and J. G. BREWEN, 1966: Cytological observations on the strandedness of Mammalian metaphase chromosomes. Cytologia (Tokyo) 31, 208—212.

URBANI, E., e S. RUSSO-CAIA, 1964: Osservazioni citochimiche e autoradiografiche sul metabolismo degli acide nucleici nella oogenesi di Dytiscus marginalis L. Rend. Ist. Sci. Univ. Camerino 5, 19—50.

VAN'T HOF, J., 1965: Relationships between mitotic cycle duration, S period duration and the average rate of DNA synthesis in the root meristem cells of several plants. Exp. Cell Res. 39, 48—58.
— 1967: RNA synthesis during G_1, S and G_2 periods of diploid and colchicine-induced tetraploid cells in the same tissue of Pisum. Exp. Cell Res. 45, 638—645.
— and A. H. SPARROW, 1963: A relationship between DNA content, nuclear volume and minimum mitotic cycle time. Proc. nat. Acad. Sci. U.S. 49, 897—902.
VASIL, I. K., 1967: Physiology and cytology of anther development. Biol. Rev. 42, 327—373.
VINCENT, W. S., 1964: The nucleolus. Proc. XI. Int. Congr. Genetics 2, 343—358.
VON BORSTEL, R. C., D. M. PRESCOTT, and F. J. BOLLUM, 1966: Incorporation of nucleotides into nuclei of fixed cells by DNA polymerase. J. Cell Biol. 29, 21—28.

WALEN, K., 1963: The pattern of DNA synthesis in the chromosomes of the marsupial Potorous tridactylis. Proc. XI. Int. Congr. Genetics 1, 106.
WALKER, G. W. R., and K.-P. TING, 1967: Effect of selenium on recombination in barley. Canad. J. Genet. Cytol. 9, 314—320.
WALKER, P. M. B., and H. B. YATES, 1952: Nuclear components of dividing cells. Proc. roy. Soc. (Lond.) B 140, 274—299.
WARNER, J. R., 1967: The species of RNA in the HeLa cell. Pp. 79—99 in "The control of Nuclear Activity". New Jersey: Prentice Hall.
WENT, H. A., 1959: Studies on the mitotic apparatus of the sea urchin by means of antigen-antibody reactions in agar. J. Biophys. Biochem. Cytol. 6, 447—455.
WHITTEN, J. M., 1965: Differential deoxyribonucleic acid replication in the giant foot-pad cells of Sarcophaga bullata. Nature (Lond.) 208, 1019—1021.
WILBUR, K. M., and N. G. ANDERSON, 1951: Studies on isolated cell components I. Nuclear isolation by differential centrifugation. Exp. Cell Res. 2, 47—57.
WILLIAMSON, D. H., 1966: Division synchrony in yeasts. Pp. 351—379 in "Synchrony in Cell Division and Growth". New York: Interscience.
WILKINS, M. H. F., 1956: Physical studies of DNA and nucleoproteins. Cold. Spr. Harb. Symp. quant. Biol. 21, 75—88.
— and J. T. RANDALL, 1953: Crystallinity in sperm heads: molecular structure of nucleoprotein in vivo. Biochem. Biophys. Acta 10, 192—193.

WIMBER, D. E., 1961: Asynchronous replication of deoxyribonucleic acid in root tip chromosomes of *Tradescantia paludosa*. Exp. Cell Res. **23**, 402—407.
— and W. PRENSKY, 1963: Autoradiography with meiotic chromosomes of the male newt (*Triturus viridescens*) using H³-thymidine. Genetics **48**, 1731—1738.
— and H. QUASTLER, 1963: A thymidine-¹⁴C and ³H double labelling technique in the study of cell proliferation in *Tradescantia* root tips. Exp. Cell Res. **30**, 8—22.
WOODARD, J. W., E. RASCH, and H. SWIFT, 1961: Nucleic acid and protein metabolism during the mitotic cycle in *Vicia faba*. J. Biophys. Biochem. Cytol. **9**, 445—462.
WOLF, U., G. FLINSPACH, R. BOHM, und S. OHNO, 1965: DNS-Reduplikationsmuster bei den Riesen-Geschlechtschromosomen von *Microtus agrestis*. Chromosoma (Berlin) **16**, 609—617.
WOLFE, S. L., 1965 a: The fine structure of isolated metaphase chromosomes. Exp. Cell Res. **37**, 45—53.
-- 1965 b: The fine structure of isolated chromosomes. J. Ultrastruct. Res. **12**, 104—112.
— and B. JOHN, 1965: The organization and ultrastructure of male meiotic chromosomes in *Oncopeltus fasciatus*. Chromosoma (Berlin) **17**, 85—103.
— and J. N. GRIM, 1967: The relationship of isolated chromosome fibers to the fibers of the embedded nucleus. J. Ultrastruct. Res. **19**, 382—397.
WOLFF, S., and H. E. LUIPPOLD, 1964: Chromosome splitting as revealed by combined X-ray and labelling experiments. Exp. Cell Res. **34**, 548—556.
WOLFSBERG, M. F., 1964: Cell Population kinetics in the epithelium of the forestomach of the mouse. Exp. Cell Res. **35**, 119—131.
WOLSTENHOLME, D. R., 1965: The distribution of RNA and DNA in salivary gland chromosomes of *Chironomus tentans* as revealed by fluorescence microscopy. Chromosoma (Berlin) **17**, 219—229.
— 1966: Electron microscope identification of the interphase chromosomes of *Amoeba proteus* and *Amoeba discoides* using autoradiography: with some notes on helices and other nuclear components. Chromosoma (Berlin) **19**, 449—468.
— and I. B. DAWID, 1967: Circular mitochondrial DNA from *Xenopus laevis* and *Rana pipiens*. Chromosoma (Berlin) **20**, 445—449.
— — and H. RISTOW, 1968: An electron microscope study of DNA molecules from *Chironomus tentans* and *Chironomus thummi*. Genetics **60**, 759—770.
WOODS, P. S., and M. V. SCHAIRER, 1959: Distribution of newly synthesized deoxyribonucleic acid in dividing chromosomes. Nature (Lond.) **183**, 303—305.

YANOFSKY, S. A., and S. SPIEGELMANN, 1963: Distinct cistrons for two ribosomal RNA components. Proc. nat. Acad. Sci. U.S. **49**, 538—544.

ZETTERBERG, A., 1966: Synthesis and accumulation of nuclear and cytoplasmic proteins during interphase in mouse fibroblasts *in vitro*. Exp. Cell Res. **42**, 500—511.
ZIMMERMAN, A., 1960: Physico-chemical analysis of the isolated mitotic apparatus. Exp. Cell Res. **20**, 529—547.
ZUBAY, G., and P. DOTY, 1959: The isolation and properties of deoxyribonucleoprotein particles containing single nucleic acid molecules. J. molec. Biol. **1**, 1—20.

Species Index

Author Index